THE SCIENCE OF
NUTRITION

THE SCIENCE OF
NUTRITION

By HENRY C. SHERMAN

MITCHILL PROFESSOR OF CHEMISTRY
COLUMBIA UNIVERSITY

COLUMBIA UNIVERSITY PRESS

NEW YORK : MORNINGSIDE HEIGHTS

To

C. A. B. S.

PREFACE

THE development of the new science of nutrition is one of the major advances of our times in its dramatically rapid series of discoveries and in its potentialities both for higher health and for what Dean Woodbridge called rendering more intelligible the world in which we live.

This book seeks to present the science of nutrition in a different way from any other. As contrasted with a textbook or reference handbook, this might be called a book *about* nutrition rather than *in* the science. Technical terms are reduced to a minimum and no specific training in science is demanded of the reader. Yet somewhat more of scientific curiosity and of intelligent appreciation of accuracy and adequacy are presupposed than in the primer-like books of which so many have been published recently.

While careful thought has been given to the sequence of topics, it is realized that some readers may prefer to make their own selections of a chapter here and there. Hence it is sought to make each chapter reasonably complete in itself even though this may involve a few instances of near-repetition.

The arrangement is also such that one may easily omit any or all of the three chapters (VIII, XIV, and XV) that may seem relatively heavy, though for some readers these chapters will give added meaning to the whole.

The aim throughout is summary and evaluation of the present status of the science of nutrition in as concise and impersonal a manner as is practicable. Space has been given to historical development only where and so far as it seemed directly helpful to the understanding of present concepts. This plan has precluded any attempt to distribute honorable mention to the many past and present workers who have made important contributions to the building of this science. Any reader who is interested in recognition of individuals should read the bibliography with at least as much care as the body of the book. Even then, however, the author feels that there should here be a notice of a kind converse to that often appearing in works of fiction; in this book all omissions of names are purely accidental results of the primary objective of conciseness. In the text a date immediately following a name indicates that a reference may be found in the Bibliography; but the latter includes also a number of selections intended rather to supplement the text than merely to support it.

H. C. S.

February 4, 1943

CONTENTS

CONTENTS

LIST OF TABLES

THE SCIENCE OF
NUTRITION

CHAPTER I

THE MODERN VIEW OF FOOD AND NUTRITION

IN THE science of nutrition, as in many other fields, the modern view is essentially dynamic and functional. The function of food is to nourish, and this it does in three main ways:

First, it furnishes the fuel to yield energy for bodily activities.

Second, it supplies the structural materials for the growth and upkeep of all the varied tissues.

Third, it provides the substances which serve to maintain the body's self-regulatory system and the physico-chemical conditions within the tissues and fluids which directly environ the processes of life.

In all these aspects of the function of food, but especially in the third, our knowledge has been so greatly advanced by recent researches that our present-day science of nutrition is essentially a twentieth-century development.

At the turn of the century the outstanding news in nutrition was that concerning "the man in the copper box." By this phrase the newswriters referred to the experiments in which the energy transformations in the human body were first studied in terms of accurate measurement. Then within the decade came the opening of the modern era of emphasis upon completeness and precision in our studies of the substances of which our bodies are composed and the quantities in which we need them in our daily food; while the second decade of the century saw much development in both these directions along with a rapid series of discoveries in the vitamin field.

Together, these developments have opened the era of what has come to be called the newer knowledge of nutrition.

While the obvious outward mark of the period of this newer knowledge is the discovery of the vitamins, it is also noteworthy that the science of nutrition is evolving a fundamentally new viewpoint. This centers in the fact that what one takes in as food, even within the range of everyday normal conditions, may influence the body's internal chemistry much more significantly than hitherto supposed.

Internal environment is the term now increasingly used by English-speaking students of science to express the twentieth-century form of Claude Bernard's concept of the *milieu interne*. The present-day view differs from the original concept in that, in the light of well-established physico-chemical laws and principles, we now recognize the possibility of consciously influencing our internal environment by our choices of food. The previous two generations had taken too literally Claude Bernard's dictum that the *fixité* of its internal environment enables the body to cope with a new or changeable external environment. The neat form of this generalization kept it current as a sort of scientific dogma which we are only just now freeing from its too fatalistic implications. If the internal environment were strictly fixed we should be unable to influence it; but we have lately learned that we can and do influence it through our nutrition. This fact, together with the enlarged opportunity it brings for personal efficiency and public well-being, the present book endeavors to explain.

The present-day view of internal environment comprises new knowledge of three kinds.

First, the above-mentioned physico-chemical principles have been so well developed and so universally accepted as laws of nature that one must expect something to happen whenever one changes the proportions of chemically active substances in a "system," whether this be the glass-enclosed system of an inorganic experiment or the bodily system of a person sitting down to a meal. Many of the constituents of our ordinary foods are of such chemical natures that a larger or smaller intake makes a difference in the body's internal chemistry.

Secondly, new and more delicate methods enable the laboratory investigator now to measure nutritionally induced chemical differences which previously could not be detected.

Finally, feeding experiments with laboratory animals have in very recent years been so developed as to permit of the accurate study of relatively small nutritional differences throughout entire life-cycles and even successive generations.

Thus the effects of what is taken into the body as nourishment can be measured both in terms of the life history and of the internal chemistry. And it is found that a difference may look small in terms of chemical analysis, yet may have profound effects upon the life process in the long run.

Here the new knowledge gained through recent research is making a far-reaching change in a fundamental scientific viewpoint, showing us that we can exercise a greater degree of conscious influence over our life processes than anyone had supposed for at least a century past.

For, just about a century ago, studies of plant and animal growth and productivity indicated that living things grow and produce only so fast and so far as is consistent with the attainment of the specific chemical composition of their kind. Thus Liebig advocated fertilization of land with phosphate in expectation of a larger crop rather than a crop containing a higher percentage of phosphorus.

The idea of a strictly specific chemical composition and correspondingly fixed internal environment has served excellently as a first approximation; but now there is need of a thorough-going revision of this view.

If chemistry and physiology stood apart from each other, one might accept the advance of physico-chemical knowledge as valid for its field, and yet believe that the self-regulatory "mechanisms" of the body would keep it in the same fixed nutritional status throughout all normal variations of food intake. In fact this is what chemists and physiologists have believed hitherto. Now, however, investigations employing more delicate methods of analysis and larger numbers of experiments, with statistical treatment of the findings, are revealing differences which previously were either unsuspected or regarded as insignificant.

Thus we here have an illustration of the way that the qualitative and quantitative phases of scientific progress interlock. The chemist's obvious sequence of first forming a qualitative acquaintance with the factors concerned in a complex process and then proceeding to measure quantitatively the influence of each often proves in actual research to constitute one cycle of a spiral progress. For the quantitative phase of a far-reaching scientific development is often so much more than merely measuring that only through its precision can science gain the deeper insights which reveal the unforeseen; so that qualitatively new concepts arise from quantitative research.

This is the present position of that whole aspect of nutritional research which centers, in functional terms, upon the body's regulatory processes and internal environment; or in terms of the nutrient intake, chiefly upon the mineral elements and vitamins.

Fully aware that it is not yet certain just how many mineral elements and vitamins are necessarily involved in the nutritional process, we nevertheless have a sufficient acquaintance with the qualitative aspects of the problem to enable us to go on to that type of research in this field which is more rigorously precise and quantitative. Thus to the qualitative question, What things are involved? we add, with reference to such of these things as show greatest probability of being determining factors, the quantitative questions, How much do different foods supply? How much does the nutritional process require? and more ambitiously, How much gives the *best* results?

And at this same time, the recent quantitative and long-term development of experimental-feeding research now gives evidence of a new order of objectivity and convincingness.

For, the well-controlled laboratory colony of properly chosen species, from which experimental animals of like hereditary and nutritional backgrounds can be drawn at precisely known ages, grouped in strict parallel, and fed the diets to be compared throughout entire life-times and successive generations, in sufficiently large numbers for conclusive statistical analysis of results, constitutes a quantitative research instrument of a kind which has never been available before.

Here nutritional research not only seeks to understand the normal processes and to restore abnormalities to normal; but considers also the more constructive question, whether by finding (and thereafter following) the optimal levels of intake of those nutritional factors which prove most influential in this connection, it may be possible to effect significant improvement in the life process by the more scientific guidance of the relative proportions in which active chemical factors are introduced into the body through its food.

So if we view the development of the science of nutrition as analogous to the working out of a complex chemical problem, the immediate significance of the discovery of vitamins and the awakening to the importance of mineral elements in nutrition was to make possible an adequate working knowledge of what things are concerned in our nutrition. Chemical research has then promptly pressed on from the qualitative detection of the essential factors toward the quantitative determination of their optimal proportions. And while at first glance this might appear just the chemist's usual progression from qualitative to quantitative, yet the precision of the quantitative phase of the research discloses something that previously could not be perceived.

For we find that a food supply which is already adequate may be so improved by enrichment in certain of its chemical factors that the normal life process is measurably bettered.

The individual results are still within the range of the normal; but the average status is improved at all stages of the life cycle; many young lives are saved; the level of positive health is raised; and adults live longer.

In what follows we shall attempt, so far as the necessity for brevity permits, to indicate both the scientific evidence and its human implications.

CHAPTER II

FOOD AS FUEL AND THE BODY
AS A MACHINE

The Energy Aspect of Nutrition

IT IS significant that Lavoisier's same experiments and writings of the latter part of the eighteenth century have secured his fame both as an outstanding modernizer of the chemistry of his time and as "father" of the science of nutrition. This is chiefly for the reasons: first, that he consistently emphasized the importance of studying the quantitative relationships in chemical phenomena; and, second, that he clarified the chemical nature of combustion and related oxidations and showed the essential similarity of the burning of a candle in the air and the oxidation of the fuel foodstuffs in the body.

No attempt can here be made to mention all the important contributors to the development of the science of nutrition. Nor should the reader assume that those whose names are omitted are less meritorious than those mentioned, for our function is not to distribute honors. On the contrary, in order to tell the scientific story as concisely as possible, we shall do it in an essentially impersonal way, using names only when they conveniently help forward the story of the scientific development, and not with any thought of assuming responsibility for equitable distribution of recognition to individuals.

Several countries and many men contributed to the development of the knowledge of foods as fuels and of the fact that the energy which the body makes manifest in the work of its muscles must in the long run be supported by the potential energy of the foodstuffs which it burns. More precise quantitative relationships in the energy values of foods and the energy requirements of the body were naturally sought, as nutrition, like other sciences, aspired to make itself more and more an exact science. Just before the turn of the century Atwater, Rosa, and Benedict of the departments of chemistry and physics of Wesleyan University, Middletown, Connecticut, had brought to completion in the basement laboratory there an apparatus of unprecedented accuracy for the quantitative study of the energy aspects of human nutrition. The central feature of this apparatus was a little room about 8 feet long and 4 feet wide which could be sealed air-tight except for the controlled current of air for ventilation and rendered impervious to gain or loss of heat except by the means provided for accurate measurements and removal of the heat produced. In this room a man would live for from 3 to 13 days, receiving prearranged amounts of food to eat and of water to drink. This multiple-walled room (the newspaper phrase "copper box" arose from the fact that the innermost wall was of copper) was so equipped with chemical and physical devices, many of which had been evolved by the research workers for this specific purpose, as to make the whole an instrument of precision of a twofold kind: a *respiration calorimeter* because it permitted simultaneous determination of (1) the chemical data of the respiratory exchange and (2) the body's output of energy measured in terms of heat units. Already the *bomb calorimeter* had been developed to high accuracy in the measurement of the energy values of foods. The chemists, physicists, and physiologists of the nineties felt little doubt of the ability of the experimental science of their day to measure the fuel values of the foodstuffs, the energy that had been kinetic in the sun's rays and had become potential in the sugars, starches, and other organic foodstuffs which plants form from carbonic acid and water under the influence of sunshine. These scientists did, however, greatly doubt whether experimentation could be so formulated as to bring the net result of all the body's intricate transformations of energy into one unified measurement of energy "expenditure," "need," or "requirement" comparable with the measurements of the energy values of foods.

It was, therefore, an epoch-marking development when, just about at the turn of the century, Atwater and Benedict were able to publish the results of repeated experiments (each of several days' duration) with men, in which quantitatively accurate accountings of the energy aspects of nutrition were made. The body's energy expenditures were measured directly in the respiration calorimeter. The energy corresponding to the chemical determination of the kinds and amounts of food-stuffs burned in the body was computed. The two sets of data showed agreement well within the most optimistic expectations as to the probable experimental error of such research. Even a single three-day experiment rarely showed a difference of more than 1 percent. In the first and second series of these experiments, totaling 117 experimental days, the total computed energy was 344,138 Calories [1] and the directly measured energy was 344,619 Calories, a difference of only 0.11 percent. When Armsby later compiled all such data available up to 1913, including four series of experiments with men and one series each with dogs and with herbivorous farm animals, a total of 336 experimental days showed for computed energy 1,441,691 Calories and for directly measured energy 1,445,398 Calories, a difference of 0.25 percent.

Both the comparisons just cited were considered as substantially exact agreements, showing that the principle of the conservation of energy holds true in the body with the same precision as in an engine. Or, to those who already considered the principle of the conservation of energy as undoubtedly holding good throughout all nature, the close agreement of these direct and indirect measurements of the energy involved

[1] This is the greater calorie (or kilogram-calorie or kilo-calorie), the amount of heat which raises the temperature of one kilogram of water through one degree centigrade.

In this book, in order to comply with the convention of the physicists concisely and still not make capital letters too common, we give the Calorie an initial capital in those cases only in which an explicitly quantitative statement is being made.

in nutrition meant an extremely important extension of exact science into physiology. Thus Armsby [2] pointed out that these results may be taken as demonstrating that the animal heat arises from the combustions in the body, but that also they have a much broader significance. They show that the transformations of chemical energy into heat and work in the animal body takes place according to the same general laws and with the same equivalencies as in our artificial motors and in lifeless matter generally.

As a result of the close agreement between the measurements of energy directly and as computed from the data of the respiratory exchange, these two modes of procedure came to be called *direct* and *indirect calorimetry* and to be regarded as interchangeable for the purposes both of medical practice and of most types of research.

While Atwater and Benedict continued active experimentation upon the energy aspects of normal human nutrition by means of their respiration calorimeter, modifications of this apparatus were built by Lusk and DuBois for the study of medical problems and by Armsby for the investigation of energy metabolism and food-utilization in farm animals. Each of these projects as well as the work of the Carnegie Nutrition Laboratory established in Boston under the directorship of Dr. Benedict has inspired other workers to enter the field. DuBois and Murlin with their respective coworkers have developed especially the medical and pediatric aspects.

The energy aspect of nutrition has perhaps more fully won the status of exact science than has any other aspect, and its future as a field of research is perhaps the most permanently provided for. Only the exceptional research problems in this field now require direct calorimetric measurements, and for such work there are in the United States at least three permanently endowed laboratories: the Carnegie Institution's Nutrition Laboratory in Boston; the Russell Sage Institute of Pathology at the New York Hospital; and the Department

[2] Armsby, 1913, *Food as body fuel.*

of Vital Economics at the University of Rochester. Analogous studies of farm animals are provided for by the Institute of Animal Nutrition at Pennsylvania State College, and the well-established coöperation between the Carnegie Nutrition Laboratory and the University of New Hampshire.

Meanwhile the acceptance of indirect calorimetry has resulted in the development of a number of forms of apparatus which are convenient to use and inexpensive both in first cost and in operation. It is this development which has made practicable the wide introduction of quantitative studies of the energy aspect of nutrition (especially the so-called *basal metabolism tests*) in medicine.

Thus in 1936 Benedict wrote in his Annual Report as Director of the Nutrition Laboratory of the Carnegie Institution: "In 1910 basal metabolism measurements could be made only in special laboratories. Today there are instruments in no less than 15,000 physicians' offices, laboratories, and hospitals where such data can be readily and accurately secured."

Undoubtedly the number of such instruments in medical use has very greatly increased since Benedict's estimate of 15,000 in 1936. Moreover, special forms of respiration apparatus have been adapted to use in the field work of anthropologists and in studying the influence of various activities upon the energy exchange or expenditure.

The main trends of this rapid extension of study of the energy aspects of our nutrition are quite clear. On the one hand, only small or uncertain racial differences have been found among people observed in different parts of the world. On the other hand, one may now speak with confidence and with considerable precision of the energy needs of normal persons of different ages, sizes, and activities, and of the directions of departure from the normal in at least the more familiar of the diseases. The effects of climate are still a subject of active research, but in general are not major influences.

The energy values of our staple foods and of the individual organic substances which they chiefly contain have also been measured quantitatively. Not quite all (but probably well over 99 percent) of the energy involved in nutrition processes comes from the fuel value of the food. A little energy in the form of light is used in the visual process and it may be this radiant energy rather than fuel energy from food which carries the impression along the optic nerve. Also, some ultraviolet light penetrates the outer layers of the skin and provides the energy for the production of vitamin D in our bodies. The above-noted differences of one or two tenths of one percent between the body's total energy exchange as measured by direct and by indirect "calorimetry" gives us an approximate idea of the relative magnitudes of all other sources of nutritional energy as compared with the fuel value of the food. Inasmuch as the total of the body's energy intakes of other kinds is well within the errors of measurement of the fuel value of the food which constitutes the one main source, we may say that within a small fraction of one percent of quantitative completeness, the source of the energy transformed in our nutritional processes has been definitely located in those constituents of the food which undergo oxidation in the body.

But the energy aspect of nutrition which we have thus briefly described is not entirely independent of the structural and regulatory aspects to be described in the following chapters. Thus, the food proteins yield digestion products which function as materials both for the building and upkeep of tissues and for the making in the body of those substances called *catalysts* which accelerate the chemical reactions by which the body makes use of its food. There is at this point a very fundamental and far-reaching interrelation between the energy and the protein aspects of nutrition.

Similarly some of the mineral elements and vitamins of food form compounds in the body which expedite the reactions that yield energy for muscular work. In fact the vitamins have been likened to the ignition sparks of the automobile engine. The energy which they *themselves* yield is negligible, but without them the parts do not function and the fuel is not properly used.

Amounts of Food Required as Fuel

In living matter, energy expenditure is incessant, but not constant in the quantitative sense. A man of average size spends about 100 Calories in an hour of sitting quietly in a comfortable chair. In an hour of sleep, his energy expenditure may be one third less; in an hour of moderate exercise, it may be threefold more.

The round figure of 100 Calories is a convenient starting point, approximating, as it does, both the energy we spend in an hour of sitting still and the energy that we get from an ordinary serving of any one of several foods. Thus a one-ounce serving of any dry cereal, or two thin slices of bread, or an average pat of butter, or an average apple, banana, or potato may be expected to approximate 100 Calories. Thus any one of these might be regarded as fuel for an hour of quiet study, or for twenty minutes of brisk exercise. Eight or nine hours of sleep, however, would require only six of these 100-Calorie portions of food.

Calculations of the energy requirements of different people, according to age, sex, size, and activity (sometimes taking account of the differing activities in each hour of the day) play a fundamental part in the technical teaching of nutrition and the practice of dietetics. For example, take the Recommended Daily Allowances published by the National Research Council in 1941 (reproduced in full in the Appendix). These provide: 3,000 Calories for the moderately active man of occupation comparable with that of a carpenter, or for the man who if his work is sedentary takes a liberal amount of active muscular exercise as recreation; 4,500 Calories for a man engaged all day in very active muscular work; 2,500 Calories for a sedentary man or a moderately active woman, women being assumed to average four fifths the body weight of men.

The allowance of 2,500 Calories for a moderately active woman or a sedentary man may also serve as a *per capita* allowance for the people of the nation. For the adults are a constantly increasing share of the population, and the more active men and the boys of 12 to 20 with their energy needs above 2,500 Calories are just about balanced by the children whose needs average below 2,500 Calories.

While the computing of diets of more or less precise energy or calorie values ("counting the calories") is an important part of student training and of the professional practice of dietetics, the normal individual need not count his calories if he watches his weight.

If the energy value of the food eaten (and digested) exceeds the energy expenditure of the body, most of this surplus food fuel is stored in the body in the form of fat. When the amount of this body fat becomes considerable, it is largely segregated as adipose tissue which (like butter) is four fifths or more actual fat interspersed with about one tenth to one fifth its weight of water and a tiny bit of protein. Thus stored fat increases the body weight by slightly more than the weight of the actual fat itself. The arithmetical relations are fairly definite. If one's consumption and assimilation of food fuel exceeds his bodily expenditure of energy by 500 Calories a day, the body weight may be expected to increase about one pound a week.

Energy values of foods are therefore of interest both *pro* and *con*. They represent the means of meeting a fundamental need of everyone's nutrition. They also indicate where to hold down if the instinctive food habit tends to lead to overweight. Overweight, especially in and after middle age, is a real health hazard. Dr. Haven Emerson has frequently quoted the rule-of-thumb "for every inch by which a man's waist-measure exceeds his chest-measure, subtract two years from his life-expectation." This is, of course, a statistical expectation, not a prediction for every individual case. One of the writer's friends is a genial human "butter ball" of well over 90; but he is a rarer specimen than a thin man of the same age.

Modern statistics confirm the old adage, "A lean horse for a long race." On the other hand, in a war-impoverished world, there may for a long time be many people for whom every possible food calorie should be conserved.

Energy values of typical foods are shown on different bases of expression in Table 1.

TABLE 1

ENERGY VALUES OF THE EDIBLE PORTION OF SOME
TYPICAL FOODS

Food	100-Calorie Portion		Calories per 100 grams	Calories per pound a	Calories per ounce
	Grams	Ounces			
Apples	156	5.5	64	300	19
Bananas	101	3.6	99	450	28
Beef	66	2.3	151	700	44
Bread	38	1.3	261	1,200	75
Broccoli	270	9.5	37	170	11
Butter	14	0.5	733	3,330	208
Cantaloupe	360	12.7	28	130	81
Carrots	224	7.9	45	200	12
Cheese, Cheddar type	25	0.9	393	1,800	113
Eggs	64	2.3	158	720	45
Grapefruit	226	8.0	44	200	12
Kale	201	7.1	50	230	14
Lettuce	549	19.4	18	85	5
Milk	146	5.2	69	313	20
Oatmeal, dry	25	0.9	396	1,800	113
Oil b	11	0.4	900	4,080	255
Oranges	199	7.0	50	227	14
Potatoes	117	4.1	85	386	24
Salmon	59	2.1	169	767	48
Tomatoes	441	15.6	23	100	6
Wheat, shredded	27	1.0	369	1,675	105

a In round numbers.

b Corn (maize), cottonseed, olive, peanut, soybean, and other salad or cooking oils all have essentially the same fuel value.

Work and Fatigue

Thus far we have used the term energy in the mechanical sense, the sense in which the science of physics uses it. While the colloquial use is naturally looser, yet at times it may connote something of higher importance than energy in the sense of the science of physics, and at times the two meanings are somewhat mingled and confused. This is even true of the common definition: "Energy is the power to do work." Here the physical concept carries over into the popular way of speaking, which has a psychological connotation not recognized by physics.

So also with our use of the word fatigue. In purely physical work of a given kind, fatigue may be a function of the amount of physical energy expended. But work involving close application may be fatiguing even though the expenditure of physical energy is small. Sitting in a conference or in classroom for an hour one may spend only about 100 Calories, yet may feel quite fatigued at the end; then one may go for an hour's walk, spending twice as much energy in the physical sense, yet returning refreshed and "rested."

It may be emphasized, however, that if fatigue or nervousness increases muscular tension the body's energy expenditure may thereby be much increased. Even when the body is most completely at rest there is still enough tension in the muscles to involve probably as large an energy expenditure as that due to the work of the heart and of breathing combined. It is because of the heat resulting as "degradation product" from such internal work that under our ordinary comfortable conditions of housing and clothing the body rarely needs to burn much of its food-fuel for the purpose of warmth alone.

Does Mental Work Increase Energy Expenditure?

While mental work can be very tiring, the expenditure of energy (in the physical sense) which it involves is extremely small. Dr. F. G. Benedict reported from experiments with himself that the extra energy exchange involved in an hour of mental work amounted to no more calories than can be supplied by half a peanut.

Previously he had tried the experiment of having several college students, one at a time, take mid-year examinations in the respiration calorimeter. Here the results were less regular, doubtless because of more complicated changes in muscular tension. One student concentrating upon the writing of his examination paper becomes tense muscularly as well as nervously, while another becomes muscularly more and more relaxed as he concentrates quietly but effectively upon his mental work. As we all have many-fold more muscle than brain, the differences in muscle tension may easily overshadow the physical effect of mental work. So some of these students spent more, and some fewer, calories on the examination day than on the control day, while the average for all the students indicated just a trifle more energy expenditure as the result of their mental activity.

The Problem of Bodily Adjustment to Different Levels of Energy Exchange

F. G. Benedict studied the effects of systematic reduction of food intake in a group of healthy young men, students in a physical-training college. When the food restrictions had reduced the body weight by 12 percent, the rate of basal energy exchange per unit of body weight was found to have been reduced by about 18 percent. The men then continued for several weeks to live on this lowered level of calorie intake and energy exchange. Benedict considered that during this time they used about one third less calories *per capita* per day than they would have used had they undergone no food restriction. The general health of these men and their ability to do their accustomed work, both mental and physical, were fully maintained; but there was an unquestionable, though not easily definable, cooling or depression of their animal spirits. Whether this would have affected their initiative or endurance in an emergency we cannot tell.

In general, we must consider that the body has but limited (or slowly developed) power to adjust its rate of energy exchange, except by alteration of its muscular activity.

Maynard and McCay have shown in experiments with rats that growth may be suspended by restriction in the amount of a diet which was so rich in its protein and vitamin content that the deficit was presumably one of food calories only, and that a temporary stunting of this kind does not necessarily preclude a subsequently normal life history. Maynard, however, has declined to give this finding any direct human application.

For the present it will doubtless be best to maintain a considerable degree of scientific skepticism both toward such expressions as irreparable injury from food restrictions and toward recommendations of fasting or undernutrition as setting the stage for subsequently superior developments in the body. Little if any clarification can be expected from the experiences of the food shortages of the conquered countries; for in these there is too apt to be the tragic combination of the "hollow hunger" of inadequate food calories, and at the same time the "hidden hunger" of mineral and vitamin deficiencies.

Criteria of Normal Fatness

Some slimming among the sedentary men is said to have resulted from wartime shortages and strict rationing of food fats in more than one European country. Such slimming may be an advantage to health in a man who has been somewhat stout; but it may be a health hazard to the woman who for the sake of style has already kept herself thin. Moreover, *overweight* is an oversimplified term. This was impressed upon examiners for the Army and Navy when it was found that the world's champion pugilist, in condition for his profession, was so heavy for his height that according to routine recruiting standards he would have been deemed too fat for military service. The discrepancy, of course, was in assuming that a weight distinctly above the average for the height is always due to extra body fat. The proper criterion of obesity is the body's specific gravity. Volume for volume, fat is lighter than water, while lean flesh is heavier, and bone is heavier still. A high ratio of weight to height if due to obesity will mean a low specific gravity, while if due to a broad heavy skeleton (as well as large muscles) it will mean a high specific gravity. So by weighing him in water as well as in air, it was duly determined that Mr. Joe Louis was physically fit for service in the armed forces of the United States.

CHAPTER III

ADVANCES IN OUR KNOWLEDGE OF THE
MATERIALS OF BODILY STRUCTURE

The Proteins and Their Amino Acids

ABOUT a century ago the Dutch chemist Mulder separated and described what he thought to be the fundamental substance of body tissues, and coined for it the name protein.[1] By derivation from the Greek, this word is a claim to priority on behalf of the material to which it was applied. This material proved to be not a pure individual substance but a mixture of a number of more or less similar substances. Thus, the details of Mulder's work were not of permanent value, but the word which he introduced has remained current as a group designation; and *the proteins,* with only moderate fluctuations of prestige, have attracted the devoted attention of a succession of able investigators whose work has supported in a sense the claim to preëminence which is implied in their group name.

This field of investigation did not have to lie fallow like that of the energy concepts initiated by Lavoisier. Other chemists could extend Mulder's work without having to await the development of a supporting concept by research in physics. During the fifty years following Mulder's publication, several different but apparently related substances were separated from plant and animal materials and described as individual proteins.

It was found that these proteins, even after the most careful separation and purification, are substances whose molecules are large and complex. Much the greater part of a typical

---

[1] Mulder's portrait and a facsimile page of his published account of this work are reproduced in Mendel, 1923, *Nutrition: the Chemistry of Life.*

protein can, however, be resolved [2] into simpler constituents, the amino acids, of which typical proteins yield from twelve to twenty-two kinds each. The constituent amino acids of the different proteins of our food and of our body tissues are thus of *relatively few kinds;* but the *relative proportions* in which they occur vary greatly from protein to protein. Hence there may be important differences in nutritive value among proteins, all of which show a general similarity of chemical nature.

In 1907, the late Dr. Thomas Burr Osborne of the Connecticut Agricultural Experiment Station, who had already done outstanding work in the descriptive chemistry of the proteins, concluded a report upon the proteins of wheat with the suggestion that among proteins which differ so widely in their amino-acid make-up it was reasonable to expect that significant nutritional differences might exist. It was not the first time that such a suggestion had been made. In fact in the preceding year Professor (later Sir) Frederick Gowland Hopkins of Cambridge University discussed the relation of one particular amino acid, tryptophane, to the nutritive values of the proteins which contain it. Osborne's work provided him with purified proteins in sufficient quantities to permit further pursuit of the problem of relations between the chemical natures and the nutritive values of proteins. With the aid of a continuous series of liberal grants from the Carnegie Institution of Washington, Osborne and Mendel carried on in New Haven an extended and epoch-making series of experiments in which different typical proteins, separated and purified by the methods which Osborne had previously developed, were fed—sometimes singly, sometimes in systematically

---

[2] In its chemical nature, the change by which a protein is resolved ("broken down") into its amino acids is an *hydrolysis.* We may thus *hydrolyze* a protein by boiling it with acid in the laboratory. Also, whenever we eat and digest protein food the protein is thus changed to its amino acids by *digestive hydrolysis.* Should the reader wish a fuller account of these topics, or the names and chemical natures of the individual amino acids, the information may be found readily in textbooks and reviews of biochemistry and the chemistry of food and nutrition. A few of these are listed in the Bibliography.

planned mixtures with each other, and sometimes supplemented by individual amino acids.

This work by Osborne and Mendel marks a turning point from the slow progress of the preceding century into the era of the modern science of nutrition. It was a distinct step into a newer knowledge of the protein aspect of nutrition, as was the work of Atwater, Rosa, and Benedict in the energy aspect, which it followed by about a decade.

Almost simultaneously with the earliest correlations of the nutritive efficiencies of proteins with their amino-acid constitution by Hopkins, by Osborne and Mendel, and by McCollum, Chittenden published his extensive studies of the quantitative requirement for protein in the normal nutrition of human adults.

Subsequently, several investigators extended both these aspects of "the protein problem" in nutrition. Particularly outstanding have become the two series of contributions: on the one hand as to the proteins contained in different natural foods as studied by D. B. Jones and coworkers in the United States Department of Agriculture; on the other hand, as to the nutritional dispensability or indispensability of individual amino acids as studied by W. C. Rose and coworkers at the University of Illinois. Both these lines of protein research have thus won the permanent interest of such able investigators, and also such adequate financial support that their further development may be regarded as well assured.

In practical problems of nutrition and food supply it would often be prohibitively cumbersome to take explicit quantitative account of each amino acid, or even of each of the ten which are now recognized as indispensable to the nutritional adequacy of a diet. So while recognizing that there are considerable differences in amino-acid make-up between *individual* proteins, we trust that these will be more or less "evened up" in the protein-mixtures of ordinary dietaries, and that it may therefore suffice for most practical purposes to continue to state the protein needs of nutrition in the simple terms of the amount of the protein mixture contained in the normal mixed diet.

For normal adult maintenance the usual allowance is about 1 gram of food protein per day per kilogram of body weight. Lewis (1942) states it as his opinion, and as the general consensus of opinion of students of the subject, that this allowance is half again to twice as high as the average of actual need. This we believe to be a sufficiently safe margin to cover individual variations (of need, and among dietaries) without complicating the "protein standard" by the added proviso (still sometimes met, but now rather antiquated) that a certain percentage of the food-protein shall be "animal protein." It is of far-reaching importance that this fact be effectively assimilated into our thinking, for in the world view of food problems, which must be accepted as one of the responsibilities of the present generation and its successors, there may not be enough animal protein to provide for the dietaries of all people as much as some of the Western groups of people have sought to incorporate in their standard of living. It should be universally recognized as *now known,* and as good citizenship to act upon the knowledge, that all scientifically sound protein standards can be met in terms of foods of which the readily potential supplies are sufficient to go around.

In his 1942 review Lewis explains how, even among proteins containing the same amino acids in the same quantitative proportions, the different ways in which these amino acids may be linked within the protein molecule make possible a variety of molecular structures extending to "a number beyond the range of human thought." This adds great emphasis to the importance of the digestive changes: first in protecting us from protein allergies, and then in sending the body tissues the amino-acid units from which the tissues reconstruct protein molecules, each organ and tissue according to its own patterns.

Here as at many other points, this essay upon the development and trend of the modern science of nutrition should take account of related biochemical advances, even though the requirements of brevity forbid any such digressions as would be involved in attempts to set them forth at all fully.

Mineral Elements

Most typical proteins are composed of the five chemical elements, carbon, hydrogen, oxygen, nitrogen, and sulfur; a few also contain phosphorus or iron, or both.

About a dozen other chemical elements are known or considered to be nutritionally essential. There is, and can be, no line of demarcation at once sharp and scientifically sound between the organic and the inorganic or mineral elements. An element may enter the body in an organic, and leave it in an inorganic or mineral, form. As a convenience we use the phrase inorganic or, more often, mineral elements for all of those which either exist in the body largely as constituents of inorganic or mineral substance, or which largely remain in the ash when a food or a body tissue is burned.

In part these elements exist in the body as soluble salts, which help to give the tissues and body fluids the properties on which the life processes largely depend. On the other hand, the relatively insoluble mineral salts of the body are built into the skeletal structures of bones and teeth.

Still a third type of occurrence of some at least of these same elements in the body is as constituents of some of the soft tissues. Thus most of the iron in the body at any given time is in an organic combination as a constituent of the hemoglobin of the red blood cells, and most of the body's sulfur is organically combined in the proteins of its various tissues. Because iron may enter and either iron or sulfur may leave the body largely in mineral form, these are commonly included among the so-called mineral elements.

Many illustrations of the functioning of these elements belong alike to general physiology and to the science of nutrition. We shall here mention, therefore, only one of them as an example. In order that the heart may continue to beat (that is, that the heart muscles may continue their orderly sequence of contraction and relaxation), these muscles must be bathed by blood and lymph containing adequate amounts and the right forms and relative proportions of sodium and calcium, both of which must in the long run be supplied by the body's nutritional intake.

CALCIUM

Sodium, calcium, potassium, and magnesium all have important relations to the maintenance of normal condition and tone in the muscles and the nervous system.

Similarly the normal solubility, transport, and permeability relationships in the body depend upon right forms and concentrations of these elements which in turn depend for their maintenance upon our nutritional intakes.

Calcium is also essential to the coagulation potentiality of the blood, without which any small bleeding would continue indefinitely; whereas with the calcium factor normal, the blood itself repairs from within the many tiny accidents which occur to its containing walls, often without our knowing anything about them.

Whether or not the *term* is "philosophically self-contradictory," the *concept* of *dynamic equilibrium* is extremely important in physico-chemical science, and not least in its clarification of some of the nutritional processes. The calcium salts of the blood and lymph are maintained within a narrow range of concentration levels (notwithstanding the varied emergencies which may make demands upon them) by virtue of the fact that they are in dynamic equilibrium with at least two forms of calcium reserve: compounds of calcium with protein in the blood plasma; and the larger though less mobile reserve of calcium existing as relatively fixed mineral tissue in the bones and teeth. Over 99 percent of the total calcium possessed by a normally developed body is thus contained in the skeletal system.

When the food shortages of the First World War and its

aftermath were faced in the dawning light of the newer knowledge of nutrition, it was realized that if the people were to
be adequately nourished, their staff of life, the grain crops,
would need supplementing with what McCollum was teaching us to call protective foods. And as the latter were more
costly the urgent question was in what proportion they must
be used to support permanently satisfactory nutrition. In an
investigation at Columbia, using white rats as the experimental
animal,[a] we found that a mixture of five parts ground whole
wheat with one part dried whole milk proved adequate to
the support of normal health, growth, and reproduction generation after generation. Yet when the proportion of protective food was higher, the life history was improved. Growth
and development were expedited, the level of stamina and
adult vitality was higher, and longevity was improved by ten
percent. Also there was an extension in still larger proportion
of the period between the attainment of full adult capacity
and the onset of senility, which for convenience we may call
the prime of life. Or in human terms, the nutritional improvement of an already normal life history included higher health
and longer life with a reduction of the period of dependency.
For brevity we may call the merely *adequate* diet A, and the
better diet B.

When diet A was enriched in calcium alone (to the level
of diet B) there resulted distinct improvements in all the
aspects of the life history above mentioned except that the
increase in length of life of the females was so small that if
considered by itself it would be insignificant. It is to be noted,
however, that these females had borne and suckled a greater
number of young. Hence the question arose whether the
females were constitutionally unable to invest extra calcium
in extra longevity or whether they had invested it in larger

[a] As these experiments were to continue through successive generations they
were necessarily made with a species whose natural life cycle is relatively short.
The rat was selected because its nutritional chemistry is so closely like the
human with respect to all the factors with which we are concerned in this
chapter.

families instead. The question was investigated in two ways, and it was found that (1) when the same calcium increment was tested with unmated females they did gain as much in longevity as did the males; and (2) when larger increments of calcium were given to mated females they were thereby enabled both to rear more and sturdier offspring and to live decidedly longer themselves.

Calcium-balance experiments with growing children of all ages have shown clearly that the level of intake of food calcium influences the rate of retention of calcium by the body and therefore the calcium content of the body at a given age.

In the animal experimentation it has been possible by direct chemical analyses of large numbers at fixed ages to establish definitely and exactly the relation of the calcium content of the food to that of the body. At the same time the continuation of the same feeding levels throughout the entire lives of parallel animals, and even through successive generations, has shown that the higher calcium contents of body which result from the higher calcium intakes are favorable to the entire life history of the individual and to the stamina of the offspring.

Inasmuch as 99 percent or more of the calcium which the body retains is held in the relatively insoluble form of the calcium phosphate minerals of the skeleton, the question naturally arises how the increased amount of so relatively segregated a mineral can have such an important influence upon the life-long well-being of the individual and even also of the next generation. An explanation is found in the bone trabeculae. These are slender crystals of bone mineral which extend from the inner surfaces of the ends of the bones toward the center of the end of the bone cavity. The higher the level of intake of food calcium, the greater the numbers and the lengths of these calcium phosphate bone trabeculae. This greater trabecular development in the calcium-rich body has both a mechanical and a chemical significance. Mechanically it means (like the bracing rods in a bridge) a strengthening of the structure of the bone joint and a more effective preservation of its precise normal form. Chemically, these trabeculae function in the maintenance of the full normal concentration of calcium in the blood. As the ends of the bones are porous and penetrated by many blood vessels, there is continuous passage of circulating blood over these trabeculae. Hence the greater the surface of calcium compound which they present to the blood, the more quickly and completely are all the many losses of blood calcium restored. Even though the fluctuations of the blood-calcium concentration are but small from the viewpoint of our ability to measure them directly; nevertheless the quicker and more completely the blood recovers from every decline in its calcium concentration the better the body maintains its highest health and efficiency.

From the viewpoint of the fact that calcium exerts, in addition to its strictly specific functions, a sort of general regulatory influence also, it is interesting to find that the "optimal plateau" for calcium intake seems to be reached a little earlier when the diet as a whole is in good all-round mineral and vitamin balance. Yet however good the dietary in other respects, the calcium intake required for best results in the long run cannot be less than about double the minimal-adequate amount. Hence in any consideration that is likely to arise in practice as to which of two levels of food calcium is better as a standard or a recommended allowance for human nutrition, one may be entirely confident that the higher is the more desirable from the standpoint of permanent well-being.

IRON AND IODINE

While the body's calcium is largely in the teeth and bones, its iron is largely in the blood. Hemoglobin, the most abundant protein of the red cells of the blood, contains one third of one percent of iron. The hemoglobin constitutes around 15 percent of the weight of the blood, and the blood around 7 percent of the weight of the body. A good average red-blooded man possesses altogether about one tenth of an ounce of iron, most of which belongs to his red blood corpuscles, while much of the remainder belongs to the muscles.

Of iron, as also of calcium, the crust of the earth has a higher concentration than we need in our bodies. With plenty of these elements literally underfoot and in the soils from which our food crops grow, the possibility of a problem of the adequacy of nutritional intake is attributable to the fact that calcium and iron chiefly occur in the soil in relatively insoluble combinations and so do not pass as abundantly into food crops as do, for instance, the more soluble compounds of some of the other mineral elements.

While the amount of iron contained in the body is small, its functions are very vital. Both figuratively and literally, the iron compounds are among the most colorful substances which the body contains. The chromatin substance of the nucleus of every cell is an iron-protein compound, and so is the hemoglobin which gives redness to blood and muscle and which has the important functions both of carrying oxygen for the reactions on which the energy aspects of nutrition depend, and of protecting the body from injury by the acids which its own metabolic processes produce.

Inasmuch as the blood, while constituting only about 7 percent of the body's weight, contains about 70 percent of its iron, and since very much the largest part of the blood iron is in the form of hemoglobin, an iron-poor person is pretty sure to show anemia of one kind or another. Much might be written about the recent studies of the different anemias; but the story would be more medical than is here our intention. Not only is the investigation of the role of iron in nutrition thus largely interwoven with pathological research, but also it is complicated by the fact that the various types of anemia differ more from each other than has often been appreciated even by the investigators in this field, so that as Whipple, one of the most outstanding workers, has written, a monument to the recent research literature of iron and anemia would be a veritable Tower of Babel.

To mention only the three types of anemia which are clearly in some sense nutritional disturbances, there are (1) hemorrhagic, (2) iron-deficiency (hypochromic), and (3) pernicious anemia.

Obviously the anemia due to blood loss lends itself readily to experimental investigation; but blood loss, of course, means a greater or less loss to the body, not only of iron, nor only of hemoglobin, nor only of red blood cells, but of all the substances contained in blood. What substance or substances will be the limiting factor or factors in the rebuilding of the blood must then depend very largely upon the absolute and relative amounts of the various blood-building substances which the experimental animal is currently receiving in his food. Whipple has therefore been careful to use a uniform diet in his extensive hemorrhagic-anemia experiments with dogs; but, as this diet is a mixture of natural foods and as we have no assurance that all the significant constituents of natural foods are yet known, the strict uniformity of Whipple's basal diet does not imply definiteness of identification of the effective substance or substances in a food which is found to be helpful to recovery from experimental hemorrhagic anemia. Whipple found liver helpful in hemorrhagic, and Minot in pernicious, anemia; but as the liver is an efficient (and notorious!) catchall for whatever circulates in the body, it did not follow that the good results in the two anemias were due to the same substance in the scientific sense of chemical identity.

Pernicious anemia is very different from that produced by loss of blood. Typically it represents a lack of one or both of two substances, neither of which enters into the formation of hemoglobin at all, but rather into the construction of the stroma or framework of the red cell in which the hemoglobin is held. Castle has shown that what is needed in pernicious anemia can be supplied by the products of the digestion of beef by normal gastric juice, though neither the beef nor the gastric juice alone will cure the typical case of pernicious anemia. Hence it is said that there is an *extrinsic factor* which must be furnished in the form of a precursor by some suitable food, and also an *intrinsic factor* which changes the extrinsic factor into the substance which actually has directly to do with the building of the structure of the red blood cells.

Thus neither hemorrhagic nor pernicious anemia is attributable to an insufficiency of food iron to meet normal nutritional needs. This latter is sometimes but not always the cause of hypochromic anemia. An experimental hypochromic anemia can be produced by iron-poor food, but it is important also to remember that as such anemias occur clinically, they may be due to faults within the body as often as (or, some physicians say, much more often than) to faulty food. No doubt both faults may contribute to the production of the anemia; but the wise nutritionist is careful to emphasize that even an anemia curable by iron is not necessarily to be blamed primarily upon the iron content of the food, for many such are to be regarded as medical rather than dietary responsibilities. It is often much better to supplement the normal dietary with an iron salt prescribed by the physician than to distort the dietary in a misguided enthusiasm for making the food furnish all the iron that the patient needs *as therapy*. It is well to discriminate therapeutic from normal nutritional responsibilities, even if they are not always different.

Iodine, unlike iron, is very unevenly distributed geographically. It occurs largely in the form of soluble iodides, accompanying the corresponding chlorides in sea water and in natural rock salts and brines. But when table salt is as highly refined as has become customary during the past generation or two, it no longer supplies us with the small amount of iodide we need, and in regions where the soils, crops, and drinking waters are exceptionally poor in iodides the population may incur a nutritional shortage of iodine, which often manifests itself as "simple goiter."

The explanation is that the iodine entering the body as iodide of food (including table salt and drinking water) is used by the thyroid gland in its production of *thyroxine,* a substance essential to the regulation of some of the life processes. When the blood is abnormally poor in iodide, the thyroid gland (as if seeking to compensate for this inadequacy of its environment) enlarges itself; and this swollen condition of the gland is called a goiter. Goiters may also result from infections. When the swelling of the gland is not attributable to an infection, it is called a simple goiter and such goiters have long been known to be much more common in some regions than in others. Such goitrous regions have now generally been accounted for on the ground that the soils, waters, and foods are too poor in iodine (iodide) to meet the needs of human nutrition. For such regions *iodized salt* is prepared by simply restoring to refined table salt some such proportion of iodide as was presumably removed in the refining. This need not be a matter of very precise calculation, for proportions of iodide varying from 1:5,000 to 1:200,000 have been used with good results. Such restoration of iodide to table salt was, in this country at least, the first publicly recognized and promoted plan of restoration of the nutritive value of a staple but artificially refined food.

OTHER NUTRITIONALLY ESSENTIAL ELEMENTS

Beside the carbon, hydrogen, oxygen, nitrogen, sulfur, phosphorus, iron, iodine, and calcium, our normal nutrition also requires potassium, sodium, magnesium, chlorine, manganese, copper, and probably cobalt and zinc. Also, there are still other elements which may turn out to play parts in our nutritional processes. When the facts are understood, it is no stumbling-block that we are not completely sure just how many elements are involved in our nutrition, for those which are in doubt are needed, if at all, only in such small amounts as we get from our environment incidentally.

Protein, Calcium, and Iron Requirements and Sources

Of the body-building nutrients which we are considering in this chapter, only three are included in the so-called "yardstick for good nutrition"—the table of Recommended Daily Allowances for Specific Nutrients published by the National Research Council in 1941 and reproduced in full in the Appendix. These allowances provide for the moderately active man (weighing, exclusive of clothing, 70 kilograms or 154 pounds): protein, 70 grams; calcium, 0.8 gram; iron, 0.012 gram. The calcium and iron allowances recommended for women are not reduced in the proportion of their lesser body weight; and for children the allowances per unit of body weight are decidedly higher.

Table 2 illustrates a few typical sources of the nutritionally needed proteins, calcium, and iron by showing the percentage of these nutrients in the same foods whose energy values have been shown in Table 1.

Those who care to read these tables, by way of becoming somewhat acquainted with the differing nutritional characteristics of different articles and types of food, will keep in mind that the numerical values given are for the edible portion of each food in the condition of wetness or dryness in which the dealer delivers it to the consuming household. If, as is sometimes done, the percentages were given on the basis of the food solids, those for fruits, fresh vegetables, and milk would be much higher, both in actual figures and relative to most other foods.

Another basis of expression of the richness of foods in specific nutrients is to state the quantities of such nutrients contained in 100-Calorie portions of the respective foods. In one respect this is quite logical; for the man needing 3,000 Calories a day will presumably eat about thirty 100-Calorie portions, however he may choose to select them. But the merit of the comparison on the basis of 100-Calorie portions may be limited by the fact that one would not eat more than a small fraction of his calories in the form of certain foods, while certain other foods may easily supply large fractions of the total energy intake.

TABLE 2

PERCENTAGES OF PROTEIN, CALCIUM, AND IRON IN THE
EDIBLE PORTION OF A FEW TYPICAL FOODS

Food	Protein percent	Calcium percent	Iron percent
Apples	0.3	0.007	0.0003
Bananas	1.2	0.008	0.0006
Beef, lean	19.7	0.013	0.0030
Bread, white	9.0	0.05 [a]	[b]
—— whole wheat	9.5	0.06 [a]	0.0030
Broccoli	3.3	0.146	0.0014
Butter	0.6	0.016	0.0002
Cantaloupe	0.6	0.017	0.0004
Carrots	1.2	0.042	0.0007
Cheese, Cheddar type	23.9	0.873	(0.001)
Eggs	12.8	0.058	0.0031
Grapefruit	0.5	0.017	0.0003
Kale	3.9	0.181	0.0025
Lettuce	1.2	0.054 [c]	0.0011 [c]
Milk	3.5	0.118	0.0002
Oatmeal, dry	14.2	0.081	0.0052
Oranges	0.9	0.043	0.0003
Peas, dry	23.8	0.073	0.0060
—— fresh, green	6.7	0.022	0.0019
Potatoes	2.0	0.013	0.0011
Salmon, canned	20.6	0.194 [d]	0.0009
Tomatoes	1.0	0.006	0.0006

[a] Varies with methods of breadmaking.
[b] Unless "enriched," white bread has only about one fourth as much iron as whole wheat. Enriched bread approaches whole wheat levels in iron, niacin, and thiamin contents.
[c] Higher in loose-leaf than in headed varieties.
[d] Including bone.

CHAPTER IV

A FRUITFUL BROADENING OF EXPERI-
MENTAL METHOD: INTRODUCING
THE VITAMINS

THROUGHOUT the early years of the twentieth century, the chemistry of food and nutrition was in an odd position. One could analyze foods with as near an approach to 100 percent as is expected with most other materials; yet one could not maintain normal nutrition with mixtures of the substances which the food analysis revealed. Moreover, the purer the substances of which such food mixtures were made, the less successful nutritionally these synthetic diets tended to be. Yet the function of food is to nourish, and the trend of scientific thinking was increasingly functional. It was time to add something new to the methods of inquiry in this field.

With the growth of the conviction that broader views and deeper insights were needed, these were sought in three main ways: (1) in the further elaboration of studies of the molecular structures of the more complex components of the body and the food; (2) in the application to nutritional problems of the new methods of physical chemistry; and (3) in the more frequent and systematic use of laboratory animals as instruments and reagents of research into food values and nutritional needs. This last development was soon rewarded with a series of discoveries of substances which are essential to nutrition but whose existence had previously been either entirely unknown or only vaguely apprehended.

These newly discovered substances came to be called *the vitamins*. They do not logically belong under any group name, for they are not sufficiently related to each other either in chemical nature or in nutritional function. The editorial writer who demanded that chemists should not add further to the list of vitamins but should, instead, discover simplifying generalizations regarding those already known, was doomed to disappointment. For the more we study the vitamins, the clearer it becomes that they are not so related as to have important properties in common. Each should be thought of, not so much as belonging to a group with other vitamins, but rather as an independent factor in food values and in nutritional needs and processes. In several cases, such as thiamin and riboflavin, a vitamin has been given a new name indicative of its chemical nature.

To emphasize that each vitamin is independent, in the sense that we should not attempt to generalize from one to another, is not inconsistent with the possibility that there may be interrelationships among them. Moreover, there is scientific significance in the fact that we owe the discovery of all these substances to the same broadening of method in the use of laboratory animals as instruments of research into the problems of human nutrition.

What appears to be the first convincingly clear statement in modern terms that adequate diet must furnish some substance, or substances, other than proteins, fats, carbohydrates, and mineral matter, was made by Professor (now Sir) Frederick Gowland Hopkins of Cambridge, England, in 1906.

Hopkins found that the addition of small amounts of milk, fresh or dried, or of an alcoholic extract either of dried milk or of certain vegetables, to diets otherwise composed of purified foodstuffs, sufficed to induce growth in young rats. Seeking further evidence as to the chemical nature of the previously unknown essential nutrient thus shown to be contained in milk and some other natural foods, but not in the highly purified foodstuffs, he deferred publication of the details of his experiments until 1912. Meanwhile, Thomas Burr Osborne and Lafayette Benedict Mendel of New Haven, Connecticut, demonstrated that a similar growth-promoting effect was obtained when they introduced into their rations of isolated foodstuffs a moderate amount of "protein-free milk" —a powder made by removing the fat and practically all of the protein from milk and evaporating the resulting clear whey. Since in both these investigations it was found that milk ash does not suffice to confer the growth-promoting property upon the mixture of purified foodstuffs, it was evident that the unidentified essential substance or substances must be organic rather than inorganic in chemical nature.

At about the same time it was made clear, largely through the work of the United States Army Medical Commission for the study of tropical diseases in the Philippines, that the nerve disease common in the Orient under the name *beriberi* is due to the lack of some essential substance existing in natural foods, but which in the milling of white rice, and of other grains to white products, is rejected in the germ and outer layers of the grain. This Army Medical Commission showed that the substance which prevented or cured the neuritis of beriberi (and which, therefore, was known as *antineuritic* substance) was soluble in water or in alcohol, that it was not volatile but was gradually destroyed by heating in solution, and that destruction occurred more rapidly when the solution was alkaline. They held it to be an organic base, but not an alkaloid. Casimir Funk, working in the light of these results, was the first to claim the isolation of this antineuritic substance. He gave to his product the name *vitamine*. Subsequent investigations make it altogether probable that Funk's product was not a pure or definite substance, but that the antineuritic substance which it contained is the same as one of the water-soluble substances in milk, to which the growth-promoting property of Osborne and Mendel's protein-free milk was due.

Almost at the same time it was found both by McCollum and Davis at the University of Wisconsin and by Osborne and Mendel at New Haven that the fat of milk (butter fat) also contains something which exerts a growth-promoting influence, and in the entire absence of which growth is impossible. Both the water-soluble and the fat-soluble substances are soluble in alcohol, which explains the fact that an alcohol extract of milk and of certain vegetables sufficed to make Hopkins' food mixtures adequate.

The name vitamine, introduced by Funk in 1911, rapidly gained currency through his writings, and the two unidentified dietary essentials came to be commonly known as the water-soluble vitamine and the fat-soluble vitamine, respectively. McCollum, having come to accept the necessity of the fat-soluble earlier than of the water-soluble substance, proposed that these substances be called "fat soluble A" and "water soluble B" until such time as their complete identification should make it possible to designate each by a scientific name indicative of its chemical nature. On the recommendation of Drummond in 1920, the two terminologies were fused so that these substances came to be called vitamins [1] A and B.

The same substance which thus came to be called vitamin B had earlier been apprehended by Eijkmann, a Dutch physician working in the East Indies, as a preventive of the nerve disease beriberi. His earliest report antedated that of Hopkins but, as an official committee later expressed it, Eijkmann presented his observations "with so strong a pharmacological bias" that their nutritional significance was not made clear until much later. Ultimately a Nobel prize award was divided evenly between Hopkins and Eijkmann as independent discoverers of the first-recognized of the substances which later came to be called vitamins.

While the discoveries of the vitamins have not dominated the development of the newer knowledge of nutrition to quite the extent that many suppose, yet they do mark a far-reaching advance both in the concepts and the research methods of medicine and nutrition. It is no exaggeration to say that recent findings in the vitamin field include a whole series of inde-

[1] In the chemists' conventional nomenclature for organic compounds, the suffix *ine* connotes a special feature of molecular structure which the suffix *in* does not. So the original word vitamine was deliberately changed to vitamin to harmonize with the wider and looser significance of the word as it became the label of a heterogeneously growing group of newly discovered substances.

pendent discoveries, any one of which would have been sufficient to give the present generation a prominent place in the history of science.

The recently acquired knowledge of the vitamins has also greatly facilitated the study of the effects of the chemistry of the food consumed upon the body's nutritional well-being over longer segments of the life cycles of higher animals than had previously been deemed amenable to this type of experimental research. Another striking feature of the advance in this field is the fact that relatively few of the discoveries can, with scientific accuracy, be attributed to an individual person or to a precise date. To a very large extent each discovery has consisted not of an isolated event, but of a gradual accumulation of evidence until finally it became convincing.

The following paragraphs of this chapter will introduce in as integrated a way as possible the significances of the chief vitamins individually and of the general vitamin concept as opening the way to a more far-reaching type of nutrition research than was previously possible. Independent short stories of individual vitamins will then follow in the next three chapters.

Vitamin A is introduced by Gove Hambidge (1939) as follows:

The retina, or seeing part of the eye, contains two kinds of structures that are sensitive to light. One kind is in the form of tiny rods and the other kind in the form of tiny cones. We "see" because these rods and cones contain light-sensitive substances, just as a photographic film is coated with a substance sensitive to light. The rods are particularly important for vision in dim light and the cones in bright light.

When light strikes these rods or cones, substances are broken down chemically, and products are formed that stimulate the nerve endings of the eye. The stimulated nerves then carry a message to the brain, and this is what we call seeing. . . .

The [light-sensitive] substance in the rods . . . is called visual purple . . . it becomes bleached in the process of chemical change, and . . . if more visual purple is to be formed . . . there must be a continual supply of vitamin A. . . .

The body makes its supply of vitamin A from a yellow substance in plant foods, called carotene. It can also get vitamin A ready-made from certain oils or fats in animal foods—butter-fat, for instance.

Then the relation of vitamin A to vision is made by Hambidge to symbolize the light brought by our new knowledge of nutrition. In a later chapter we shall consider other important functions which this vitamin also serves.

Vitamin B has been differentiated into thiamin, riboflavin, niacin (nicotinic acid), pantothenic acid, and pyridoxine, all now structurally identified, while still other substances are sometimes regarded as belonging also to this group. *Thiamin* prevents and cures some of the most prevalent of the nerve diseases both of the Orient and of our Western World; and as it aids in the nutritional chemistry of all of our organs and tissues, it is proving helpful in a surprising diversity of ills. *Niacin* has been found dramatically potent in the cure of the conspicuous inflammation of the skin (and tongue) which gives the name to the disease *pellagra* which has been extremely prevalent in our Southern states. Few discoveries could be more striking than that of the potency of this inexpensive substance in the prevention and cure of such a scourge. Yet it remains to be said that when the typical pellagrin has been cured of his pellagra by means of niacin (nicotinic acid) alone, he needs something more to make him a fully healthy man. The previous diet of the poor pellagrin has usually contained so little of foods other than grain products, fats, and sweets as to make his bodily condition that of a multiple nutritional deficiency instead of a single or simple one. Clinical treatment with the pure vitamins, separately and in combination, shows that the typical pellagrin usually needs *riboflavin* almost as much as he needs niacin, and often needs thiamin also. Moreover, in only less degree, his "one-sided" food supply is likely to have involved other nutritional shortages as well as these three. Good diet cures all of these deficiencies at once, and renders unnecessary the further investigation of the frequency in the pellagrous population of shortages other than those of niacin, riboflavin, and thiamin.

Vitamin C prevents scurvy, but if we were perfectly sure of never meeting scurvy again, we still should desire to be well acquainted with vitamin C (ascorbic acid) because it serves the body's health and efficiency so importantly in so many ways. A separate chapter is given to this important factor in our nutrition.

The vitamins D prevent rickets, recently the most prevalent disease of the temperate zone. These are fat-soluble substances to which we shall recur, following the fuller discussion of vitamin A in a subsequent chapter.

Just as the vitamins are not a naturally related group of substances, so also there is no scientific reason for arranging them in the order of the alphabetic designations which they owe to the largely accidental chronology of their discovery or recognition. In the following chapters, therefore, we shall take up the chief vitamins in such sequence and groupings as seem best suited to making an orderly presentation of the nutritional significance of each, and of the more important interrelations among them.

CHAPTER V

SHORT STORY OF SCURVY AND VITAMIN C
(ASCORBIC ACID)

SCURVY, we are told, was originally a folk word which upon adoption into medical literature was formalized as *scorbutus*. Hippocrates reported that soldiers had been known to suffer from a mysterious illness characterized by pains in the legs and gangrene of the gums, the latter sometimes going so far as to cause loss of teeth. It is supposed that this was scurvy and that the disease was not then known among Greek civilians.

Later, and farther north, scurvy was all too familiar. Yet if we try to trace its history back from the period in which it bulked largest in what there then was of European medical literature, the chief result is a renewed realization of how recent modern medicine is, and how indifferently fatalistic was the attitude toward disease of our ancestors of four or five centuries ago. In southern Europe at that time scurvy was chiefly a memory associated with the afflictions of the Crusaders. In northern Europe it was so much a matter of course in the long winter and early spring that a medical writer proposed to consider all other diseases as modifications and outgrowths of scurvy. True, these winters often involved either cruel exposure to severe weather or extreme stagnation in congested and insanitary housing, yet even so it is hard for us to imagine such innocence of science as could fail to connect the annual prevalence of scorbutic ills more clearly and definitely with the deprivations of the winter diet. At least among our British ancestors, gardening was undeveloped and even the storage for winter of such fruits and vegetables as the summer season afforded was a neglected art. It is recorded that when Catherine of Aragon, recently come to England as the bride of Henry the Eighth, wanted a salad it was found necessary to dispatch the royal gardener to Holland to obtain the materials!

Quaint as this incident may sound, we are not so far removed as we may be inclined to assume from shortages of the antiscorbutic vitamin. Only about fifty years ago, an intelligent country doctor in northern Virginia, who ordinarily made his rounds on horseback, said that he himself, "like most people" in the community, was "generally more or less troubled with rheumatism in the winter and spring." His custom was to ignore it until his joints became sufficiently sore to trouble him seriously in getting on his horse; then he sucked lemons until the soreness of his joints went away. What he called "rheumatism" was almost certainly incipient scurvy; and it is probable that it was common before the general use of canned tomatoes in those parts of this country where fruits and vegetables were not abundantly stored for winter use. Historically it is unquestionable that as potato culture became prevalent in northern Europe, scurvy became less common, and that the last great epidemic of scurvy in Ireland was a sequel of the failure of the potato crop.

Observation in Petrograd in 1917 of considerable numbers of patients with mild scurvy confirmed the belief that latent or subacute scurvy—due not to absence but to too low an intake of vitamin C—is the true explanation of much of what has been vaguely called "rheumatism" by the common people of many country districts and more or less fatalistically accepted as an affliction to be expected in the late winter and early spring. A more adequate appreciation of fruits and vegetables as food and the more general availability of these foods throughout the year has done much to forestall this annual handicap; but still further progress in the same direction will undoubtedly be beneficial to health and efficiency.

From the viewpoint of today it is clearly apparent that the handicap of scurvy delayed the spread of European civilization, particularly when exploration or colonization involved the food restrictions of long voyages or of wintering in northern regions. Thus Vasco da Gama in his voyage around the Cape of Good Hope lost more than half of his crew by death from scurvy. Jacques Cartier, obliged to winter in Canada in 1535, lost a quarter of his men and found nearly all of the others more or less incapacitated by scurvy, until friendly natives taught them that decoctions of the leaves and twigs of certain trees would cure and prevent this then mysterious disease.

The superior success of Captain Cook as an explorer was largely due to his practice of paying the same close attention to fresh fruits as to drinking water in the provisioning of his ships. Every visit to a new or promising shore had as the two fixed points of its program the bringing back to shipboard of supplies of fresh water and fresh fruit.

An instance of advanced intelligence, though handicapped by difficult climatic conditions, may be claimed for the early settlers of New England. To meet the lack of fruit they tried wine and, as a substitute for this, a crude home brew of malted grain, not clarified as in the making of commercial beer. Grains that were sprouted to just the right degree and the untreated infusions of such malted grains were found to constitute a practical preventive of scurvy, so that arrangements for the making and care of such primitive brew became a part of the plan of each party of colonists; Dr. John Nichols has described how his readings of early colonial records reveal that John Alden of literary fame was originally recruited as ship's cooper to keep the beer barrels in repair. (Subsequently, more sophisticated and less intelligent consumers were taught to prefer a clear beer—and the clarification destroyed the vitamin value.)

According to recent medical writers, Captain Lancaster of the *Dragon,* one of four ships which sailed from England in 1600 as the first expedition of the newly chartered East India Company, appears to have been the outstanding pioneer in the conscious prevention of scurvy. He alone of the four captains who sailed simultaneously, had sought to forestall "the Plague of the Sea" by including lemon juice in his ship's stores and giving three spoonfuls to each man every day. His men remained healthy and, when they arrived at Table Bay, rowed to shore not only themselves but also the other three crews, not one of which could muster enough healthy men to row a small boat, so universally had they suffered from scurvy.

A century and a half later, the British naval surgeon Lind, treating an outbreak of scurvy on board the *Salisbury,* gave his limited supply of oranges and lemons to certain men, while to other men he gave other treatments, all of which were advocated by different medical men of that time. This human experience thus partook of the nature of a controlled experiment. The men who were given the oranges and lemons recovered promptly and completely from their scurvy, and those receiving cider improved, while the other men worsened although given different acid substances which had been more or less advocated as possible antiscorbutics.

About ninety years after Lind, the American physician Budd definitely advanced the idea of an antiscorbutic property possessed by some foods to the clearer and more explicit concept of a definite substance, a chemical individual. Of this substance Budd wrote, "it is hardly too sanguine to state, [it] will be discovered by organic chemistry or the experiments of physiologists in a not far distant future."

And again about ninety years later, by the use of the methods both of organic chemistry and of experimental physiology, Budd's prediction of the chemical identification of the antiscorbutic substance was fulfilled by C. G. King, then professor in the University of Pittsburgh, now scientific director of The Nutrition Foundation.

Between the prediction of this discovery by Budd and its accomplishment by King, much was learned both from human experience and from controlled experimentation with laboratory animals. Scurvy was eradicated from the British merchant marine (as it already had been from the British Navy) by the uniform issue of lemon juice to all sailors more than ten days at sea. It was through a confusion of the juices of lemons and limes that British sailors came to be called "limeys." On land also, fruits and vegetables came into more abundant use and were more generally preserved for year-round consumption.

War continued, however, to take toll through scurvy. In the War between the States, and in the Franco-Prussian War, many people died of scurvy, where in peacetime it had been almost eradicated. During the siege of Port Arthur in the Russo-Japanese War "half of the garrison of 17,000 men" developed scurvy; and it was reported that 76,000 cases of scurvy occurred in the Russian Army during one year (1916) of the First World War.

Recent surveys in this country, particularly but not exclusively in Maine, indicate that incipient or subclinical scurvy has not been as completely eradicated as we had supposed, and make it certain that many of our people still have less than optimal intakes of vitamin C.

When chemically identified vitamin C proved to be a substance related to the simple sugars, but having a certain peculiarity of molecular structure which gives it its formal scientific name (2, 3 dienol-l-gulonic acid lactone [1])—a name too long for everyday use. Usually in naming a substance of known molecular structure either a name fully indicating the chemical nature or a contraction of such a name is adopted. In the case of vitamin C, however, the history and motive of the research which disclosed its chemical nature had been so identified with the study of scurvy that the name *ascorbic acid* for the preventive substance was adopted as historically justified and as an aid to memory. It was, of course, a convenience to identify the substance through its common name with its then most prominent significance, but in future it may seem to have

[1] Those desiring to inspect the structural formulas of vitamins discussed in this book may find them in the current editions of textbooks of organic and biological chemistry or of the writer's *Chemistry of Food and Nutrition*.

been short-sighted. For when scurvy shall have become ancient history, this substance will still be important. With this in mind, and also to minimize monotony, we here make about equal use of the names vitamin C and ascorbic acid.

Since the chemical identification of this substance has been promptly followed by its abundant production at moderate cost, experimental research has been greatly facilitated. It has been possible to arrange experiments of many kinds with the pure vitamin C as the sole variable, and thus to gain more certain knowledge than was previously possible regarding the physiological behavior of the substance in the body and the pathology which results from deprivation of it.

Of the ascorbic acid which the food brings into the body, a part is destroyed (consumed) in the tissues and a part is excreted in the urine. When the intake is diminished the urinary excretion diminishes also, but not necessarily to the same extent, so that some small excretion is to be expected even if the intake were reduced to zero. Evidently whenever the combined excretion from and destruction in the body exceeds the intake, the bodily reserve is being depleted. On feeding the vitamin after such depletion, the intake is used first for the immediate needs of the tissue activities and then to replace that which had been depleted from bodily stores. After this, if the intake affords a further surplus it will go to increase the urinary excretion. Thus the larger the intake of the vitamin and the more nearly the body is saturated with it, the larger the percentage of intake which reappears in the urine.

L. J. Harris and his coworkers at Cambridge, England, were very active in developing this method of experimentation, and in using it to determine the intake required to maintain a given degree of saturation or level of vitamin C concentration in the body, as judged by the 24-hour urinary excretion. Repeatedly they found that the onset of a common cold interfered with the smooth outcome of such a "balance" experiment by causing a decrease in urinary excretion, doubtless because of increased destruction of the vitamin in the body.

Other respiratory infections and conspicuously pulmonary tuberculosis have also been found to increase the rate of destruction of the vitamin and therefore the needed intake; as have also rheumatic fever, rheumatoid arthritis, and presumably many other diseases. Moreover, not only diseases but also physical stresses and even muscular exertion have been observed to decrease the urinary output. This is doubtless because the rate of destruction is increased and the amount needed by the body to meet its rate of wastage and to maintain its normal reserve of vitamin C is correspondingly increased.

In the wording of the preceding paragraphs, as in the earlier experimental work of this general type, differences in the level of concentration of the vitamin in the body were largely inferred. Such inference has, however, been frequently verified and largely replaced by quantitative determination of vitamin C in the blood plasma, and sometimes also in whole blood, or in cerebrospinal fluid, or in special cases in one or more of the solid tissues. The term "saturation" is used in this connection to indicate the maximum concentration of the vitamin that can be maintained in the body or in a given tissue or fluid for a significant length of time. Saturation of blood plasma in this sense has been reported at about 1.5 mg. of ascorbic acid per 100 ml. (milliliters, cubic centimeters).

We may now enquire a little more specifically as to how vitamin C functions in the body, and what happens when the food fails to furnish a sufficient amount. Here as with other nutritional essentials the physiology of the nutrient and the pathology of the deficiency state throw so much light upon each other that their study is interwoven. Furthermore, the clarification of both the physiology and the pathology depended largely upon the availability of the pure substance for use in experimental research.

While the chemical characteristics of vitamin C promise other developments from further research, its physiologically outstanding function thus far lies in its relation to the intercellular "cement substance." In George Gray's excellent exposition, the pathologists' use of the term cement and the microscopic appearance of a typical intercellular tissue are brought into analogy with the reinforced concrete of engineering construction. This analogy needs the very important qualification that the anatomical cement substance is a plastic, flexible enough to be part of a working tissue yet firm enough to keep in their right positions all the individual cells of which a bodily organ is composed.

Failure of adequate cement substance accounts both for the general weakness which is one of the characteristics of scurvy, and for the abnormal readiness with which the blood leaks through the walls of the blood vessels, thus causing the little blood spots (petechiae) which may appear in the skin of the scurvy patient, or of the blood-shot gums so often shown in the traditional medical descriptions of the disease, or the invisible but painful affliction of one or more joints (often miscalled "rheumatism"), or the hemorrhages into or through the walls of the alimentary tract causing disturbances of appetite or digestion or both, and, in severe cases, bloody diarrhea. Moreover, a shortage of vitamin C and resulting failure or even chronic sluggishness of bodily ability to form intercellular tissue and keep it in good firm condition may, as Wolbach of Harvard has especially emphasized, be the cause of profound changes in the gums and the structure of the teeth, of deformities of the bones or their lack of coördination through weakness of supporting cartilage, of degeneration of bone matrix with consequent losses of calcium, and likewise degeneration of muscles, of blood-forming tissues, and sometimes of sex organs.

When we find from the modern pathology of vitamin C deficiency how this may result in a breakdown of almost whatever bodily organ or tissue is least strong in the individual person, and when we realize further that every such lesion increases the probability of infection (or the seriousness of a low-grade infection already present), we can feel some sympathy with that North European writer of some centuries ago who

proposed to consider all diseases as modifications or outgrowths of the scorbutic condition.

Conversely, as Crandon and his coworkers in the Harvard
Medical School have shown, a young adult human body with
no weaknesses in any of its parts may endure deprivation of
vitamin C for a relatively long time before scurvy lesions appear. One such young man lived on a diet practically devoid
of vitamin C and thus reduced the amount in his blood plasma
to zero in about two months; yet his body did not show visible
signs of scurvy until the 134th day of the deficient diet, nor
distinct petechiae with retarded healing of an experimental
wound until the 161st day. Then pure vitamin C was fed; the
petechiae disappeared, the wound healed, and there was a
rapid recovery of strength, which had greatly declined during
the five months on diet deficient in vitamin C.

Here the disease experimentally induced by uncomplicated
vitamin C deficiency was typical scurvy. That it required an
unexpectedly long time to develop is probably explained by
the fact that infections or other unfavorable influences have
usually hurried the onset of scurvy in clinical cases. King and
coworkers have definitely shown that extra vitamin C helps
the body to resist some of the toxins of infectious diseases.

In view of the many conditions that may increase the body's
need for vitamin C, it is now believed best to keep ourselves
saturated or nearly saturated with this important substance.
The general view of investigators in this field is that each person utilizes to advantage, even if in part as a sort of insurance,
the amount of vitamin C which his body requires to keep itself
at or near saturation. This amount is usually about 75 to 100
milligrams a day for a healthy adult. (Table 3 shows amounts
furnished by some typical foods.) The National Research
Council's Recommended Daily Allowances may be found in
full in the Appendix.

The hitherto firmly held idea of a *fixité* of our internal environment evidently must be essentially revised when we find
that our food habits affect our body content of even a sub-

stance so soluble, so diffusible, and so readily destroyed as is vitamin C.

While as yet there have been no published researches into effects of different levels of vitamin C intake upon the complete life history, there seems no reason to doubt and much reason to believe that liberal levels of intake are permanently beneficial. In fact the benefit may be greatest at the more advanced ages which have only recently received systematic attention from investigators. The decline in vitamin C content of human tissues after the age of 45 which was such a striking feature in the findings of King and coworkers probably indicates an increased rate of destruction or wastage of this vitamin in older people and that they should have vitamin-C-rich food at frequent intervals.

The vitamin C content of foods is therefore from several viewpoints an important factor in nutritive value for people of all ages. Table 3 shows such data for approximately the same foods whose energy values and protein, calcium, and iron contents have been given in Tables 1 and 2.

Comparison of Tables 1 to 3 shows at once how strikingly different may be the relative importance of one food among others, according to whether the foods are regarded as sources of energy, of one or another tissue-building material, or of vitamin C.

Meats are not important sources of this vitamin in our ordinary food supplies, nor are eggs; nor is milk except when it constitutes a large part of the dietary. In Table 3, lean beef (which has been chosen to represent lean meats in general in these tables) is said to have a variable vitamin C content; for while it is always low as compared with most fruits and vegetables, it may be sufficient so that explorers, hunters, or soldiers can ward off scurvy for some time by eating liberal quantities of fresh meat. Impressed by this fact some explorers have thought that nutritionists must be mistaken, either in counting vitamin C so important a need or in not counting meat a more prominent source. Here as in some other cases of

TABLE 3

VITAMIN C (ASCORBIC ACID) CONTENTS OF SOME TYPICAL FOODS

Food	Milligrams per 100 g.	Milligrams per ounce
Apples, medium varieties	7.0 [a]	2.0
Bananas	7.6	2.15
Beef, lean	Variable, as explained in text	
Bread, white	traces	traces
—— whole wheat	traces	traces
Butter	traces	traces
Cabbage, raw	35. [a]	10.
Cantaloupe	29.	8.
Carrots, raw	4.0	1.1
Eggs	traces	traces
Grapefruit	39.	11.
Kale, raw	50.–100.	14.–28.
Lettuce	13.5	3.8
Milk	2.2	0.6
Oatmeal, dry	—	—
Oranges	54.	15.
Peas, dry	—	—
—— fresh, green	22.	6.2
Potatoes, raw	12.6	3.6
Salmon, canned	traces	traces
Strawberries	34.	9.6
Tomatoes	23.	6.5
Watermelon	6.6	1.9

[a] Each figure stated to the presumably significant number of places.

apparently conflicting views, each of the specialists is right in his own sphere. An explorer or an army, lacking food of higher antiscorbutic value, may be saved from scurvy by large use of meat eaten immediately upon killing and including the organs as well as the muscles. But the public addressed by the nutritionist does not have access to the same sort of meat supply, for the meat of the retail market has been an unknown length of time in storage and even after delivery to the consumer will probably reach the table only after more delay and much more cooking than the fresh kill in a camp. So while frontiersmen and even an occasional army may depend upon its meat supply for its vitamin C, we usually do much better to look to the vegetable kingdom for our supply of this particular vitamin.

Mature "resting" seeds contain little if any ascorbic acid, but some is formed when the seed germinates. The sprouting of seeds may therefore be utilized as an emergency means of obtaining an antiscorbutic food. The Chinese custom of using germinating beans or peas in some of their popular dishes has undoubtedly the nutritional merit of materially increasing the vitamin C intake. Russian folklore also attributes antiscorbutic value to their slow, sour fermentation whole-grain bread; but whether this means a significant production of vitamin C in fermentation as in germination, or is only an association of the ideas of sourness and of antiscorbutic value is uncertain.

We may be sure, however, of the antiscorbutic significance of the Russian habit, described by Tolstoy in *War and Peace,* of scouring the open country in the spring and devouring almost all kinds of sprouting vegetation. Similarly, wild onions are known to have saved American frontiersmen from scurvy in pioneering times. For practical purposes under our ordinary conditions we do not count on getting any vitamin C from mature seeds (represented in Table 3 by oatmeal and dry peas), and only such traces from bread as are attributable to the milk used in breadmaking.

Our richest common sources of vitamin C are citrus fruits, tomatoes, raw cabbage, and greens of the cabbage group such as broccoli and kale. American farmers appreciate kale in the feeding of their cows and chickens; but American families have tended to neglect it in feeding themselves and their children, although the "kailyard" is often in the domestic landscape of English and Scottish literature. It is reported that the Second World War has brought a marked revival of kale culture in Great Britain.

Cantaloupes and strawberries are, at least when in season and eaten fresh, almost as rich in ascorbic acid as the foods mentioned in the preceding paragraph. Such foods as apples, bananas, and potatoes, while distinctly less rich, weight for weight, may yet be important sources of vitamin C because of the quantities in which they may readily be consumed. (Cooking losses are reported as varying widely in the case of potatoes.)

Because of our new knowledge of the importance of vitamin C and because, by reason of its labile nature, much of it disappears rather rapidly from the blood, we now believe it a good investment to include some significant source of this vitamin in every meal, and to provide fruit or fruit juice instead of sweets if snacks are to be eaten between meals or at bedtime. The science of nutrition thus brings us the pleasure of knowing that such liberal consumption of fruit is not an extravagance but a sound investment.

In conclusion, it may be said that, while scurvy as a scientific problem has been solved, the story of vitamin C is undoubtedly to be continued through further research. Much remains to be learned as to the various ways in which this substance may function in the different chemical processes which go on in our bodies and as to how the life process as a whole is influenced by habitually higher or lower levels of vitamin C intake. While any attempt to foretell the outcome of a scientific research is "the most gratuitous form of error," it is also true that guiding hypotheses are helpful in stimulating research. Such an hypothesis suggests that liberality of intake of this vitamin may be a factor in what McCollum and Simmonds have attractively called the conservation of the characteristics of youth.

CHAPTER VI

THE OUTSTANDING VITAMINS OF THE B GROUP

Thiamin (Vitamin B or B₁)

WHEN America became responsible for the Philippines, the new officers who took over the administration of the large Bilibid prison at Manila were moved with pity to find that the prisoners subsisted largely upon rice and that the rice provided for them was of the lowest commercial grade. This they replaced by high-grade white rice. They also introduced sanitary improvements and sought to make their administration in all respects humane. Yet a few months later they found themselves confronted with an epidemic of the Oriental disease *beriberi*. Since Pasteur, epidemic disease had come to be regarded as meaning (or involving) infection even if the infective agent were not yet known; so sanitary improvements were pressed still further. Yet the epidemic of beriberi continued. American army medical men, studying the literature of this baffling disease, found that there was also a nutritional hypothesis as to the cause and nature of beriberi. They then again changed the prison ration, reducing the predominance of rice and increasing the allowance of foods reported beneficial in the prevention and cure of this disease. Soon the number of cases of beriberi began to decrease and within four months the prison was freed from this year-long epidemic.

Army medical officers also found that the body of native troops known as "the Philippine Scouts" suffered much from beriberi, and that the disease could be eradicated by the addition of beans to the ration (if all the soldiers could be persuaded to eat them) or by the use of brown undermilled, instead of white polished rice.

By this time an American Army Medical Commission was systematically engaged in the investigation of beriberi and the other Oriental and tropical diseases of the Philippines; and European physicians in the East Indian services of their respective countries were similarly engaged. What had previously been a neglected nutritional hypothesis now became the accepted theory of beriberi, and search for the substance having power to prevent and cure the disease was carried on actively at several different research centers. Beriberi was known chiefly as a nerve disease, its formal characterization being *a multiple peripheral neuritis*. Hence its nutritionally postulated preventive was called *the antineuritic substance*. As we saw in Chapter IV, the idea of such a substance was more or less implicit in the papers published by the Dutch physicians Eijkmann and Grijns about 1906. It was much more explicitly developed by the British investigators Frazier and Stanton, and by Vedder and Williams of the American Army Medical Commission in the Philippines. Each of these research groups made good progress toward the isolation of the antineuritic substance and learned many of its chemical, physical, and physiological properties. Funk, a Polish chemist working in the laboratories of the Lister Institute of London, extended the isolation experiments just mentioned; and when he had separated what he thought to be a sufficiently pure preparation of the antineuritic substance, he coined for it, as we have noted, the name *vitamine, amine* being the chemical name for the group of substances to which this was thought to belong. Thus the name was introduced (near the end of 1911) after the concept was already sufficiently developed to be the guiding principle of several laboratory, clinical, and field researches.

In one of these field researches, 300 laborers engaged on construction work were encamped in a virgin jungle, under exactly the same housing and other conditions except that one half of them received unpolished brown rice and the other half polished white rice. Beriberi developed in a large proportion of those on white rice and in none of those subsisting upon rice which had not been thus impoverished by milling.

Simple extracts of the rice polishings (the germ and outer layers of the grain) were found to have preventive and curative value against beriberi. Dr. R. R. Williams, an American chemist then on the staff of the Philippine Bureau of Science, had joined in the researches initiated by the Army Medical Commission following the experiences at Bilibid prison and with the Philippine Scouts. In the civilian population, beriberi appeared frequently among lactating mothers and often affected their babies as well. In cases which had not become too severe, rice-polish extract given to the mother cured both her and her nursing child. Williams was thus able to demonstrate directly with human subjects the progress of his researches upon the laboratory extraction and approach toward isolation of the antineuritic substance, also called antiberiberi vitamine or vitamin B.

In the hope of better facilities for the laboratory isolation and chemical identification of this vitamin, Williams transferred from the Bureau of Science at Manila to the Federal Bureau of Chemistry at Washington, where the present writer was privileged to talk with him about his work in 1916. Then the First World War claimed his services for researches of more immediately military significance; these led to responsibilities in an industrial research laboratory which have continued to claim him ever since. In his spare time, however, he continued his investigation of the chemical nature of this vitamin. His original discoveries made possible the isolation of the pure substance from natural sources in quantities sufficient for effective studies of its molecular constitution, which was finally established in 1936. Very soon after this the synthesis of the substance also was accomplished and industrial manufacture of the pure vitamin at moderate cost was made possible.

In several phases of his research, Williams enjoyed, as he has repeatedly explained in generous terms, the partnership of able coworkers. Other investigators were independently active in the study of the same substance at the same time, but science is especially indebted to Williams and has shown recognition in many ways, including the award of the Gibbs Medal of the American Chemical Society and the Chandler Medal of Columbia University.

With the chemical identification of the substance, it received, of course, a systematic chemical name indicative of its full molecular structure; but this formal name is far too long for use as an everyday designation, so the name *thiamin* [1] or *thiamine* was devised to suggest as much of the chemical nature as can be told by a single word of moderate length. The pure substance is ordinarily handled as the salt which it forms with hydrochloric acid, so this crystalline material is called thiamine chloride or thiamine hydrochloride. There is no serious scientific or practical objection to the common practice of using the single word thiamin or thiamine to designate both the substance itself in whatever form it may occur in foods and the salt which constitutes the crystalline synthetic substance of commerce.

Brief mention has already been made of the fact that the existence of the substance we now call thiamine was discovered in two quite distinct ways: through laboratory research in normal nutrition, and through clinical and field studies of the nerve disease beriberi.

Combining the findings of both types of investigation we may now summarize the outstanding relations of the substance to nutrition and health.

It is essential to the growth of the young, and to the maintenance of normal appetite at all ages. The latter function is something more than merely making the food appetizing: it

[1] There are reasons both pro and con the final e of thiamine *versus* thiamin. The reader should be accustomed (and undisturbed) to meet this word in both these forms—also as thiamine hydrochloride.

is an effect of the vitamin upon appetite as a physiological function. When experimental animals are supplied *ad libitum* with a food mixture good in other respects but lacking thiamine, after a while appetite fails. If they then are fed thiamine *even quite separately*, its effect is such that they return with appetite to the same food which they had refused. Along with the loss of appetite there is often a loss of gastrointestinal tone and a general weakening of the body. All these pathological effects of a lack of thiamine may be more or less definitely referable to the effect of this specific kind of starvation upon the process through which the body utilizes the greater part of its food-fuel, namely the sequence of chemical changes through which these fuel foodstuffs, more especially the carbohydrates, are made ready for the oxidation involved in the energy aspect of nutrition. In general, every active cell in the body is its own energy transformer. Hence, whatever organ or type of tissue is least resistant may feel the depression involved in any such handicapping of the normal energy relationships, and conversely if already so depressed (whatever the cause) may show benefit from an increased thiamin intake. So thiamine has been found helpful in the treatment of a varied list of ills. And so, too, it is thought that the failures of appetite, of gastrointestinal stamina, and of general bodily tone may be results of the handicapping of an essential step in the fundamental chemistry of the body's nutritional process.

From this concept it is, of course, a simple and logical step to the question whether less drastic shortages of thiamine, such as might never develop the full-blown symptoms of beriberi, may not be responsible for less clear-cut but perhaps much more prevalent ills.

Especially influential in this connection are the findings of Drs. R. M. Wilder, R. D. Williams, and their coworkers. Under their medical guidance and observation, about a dozen young women volunteered as subjects in studies of the effects of long-continued, low-thiamine diets. In addition to the analytically determined effect of the shortage of thiamine upon the bodily chemistry, these women showed, as the results of a long-continued slight-to-moderate shortage of thiamine, such changes of behavior as would justify a diagnosis of neurasthenia "and other objective evidence of psychic disturbances involving mental depression, irritability, and loss of efficiency." This is strong evidence from an eminent medical source that a higher intake of thiamine might mean a lower proportion of neurasthenics among our American people. Coming at the time in our national history that it did, its colloquial interpretation was to dub thiamine "the morale vitamin." This, of course, is an oversimplification. The undoubtedly important function of nutritional well-being and the resulting positive or buoyant health in promoting morale depends upon appropriately liberal intakes of several nutritional factors rather than of any one alone. Yet it is possible that thiamine may have a somewhat outstanding influence upon the stability of the nervous system, as it seems to have upon the stabilization of appetite.

The work of Wilder, Williams, and others led them to the conclusion that the amount of thiamine needed for *optimal* nutrition is not less than 0.5 milligram nor more than 1.0 milligram per 1,000 Calories obtained from a diet of ordinary composition. This, it will be noted, is the opinion of expert physicians as to the level of intake that can be expected to yield the very best results. It is considerably above the level which statistical evidence indicates as necessary for the prevention of beriberi.

Table 4 shows approximate thiamine contents of about the same foods used as illustrations in previous tables.

Beginning in 1940, Dr. Wilder, Dr. Williams, and others have consistently and effectively urged that white bread should be enriched with thiamine to offset, at least in part, the impoverishment of the American dietary which had occurred during the past two generations, largely through the use of highly refined flour in breadmaking. While Wilder and Wil-

TABLE 4

THIAMINE CONTENTS OF SOME TYPICAL FOODS
(EDIBLE PORTION)

Food	Milligrams per 100 g. [a]	Milligrams per ounce [a]
Apples	0.038	0.011
Bananas	0.075	0.021
Beef, lean	0.160	0.045
Bread, white, not enriched	0.07 [b]	0.02 [b]
—— white, enriched	0.25 [b]	0.07 [b]
—— whole wheat	0.35 [b]	0.10 [b]
Butter	trace	trace
Cabbage	0.105	0.030
Cantaloupe	0.058	0.016
Carrots	0.100	0.028
Eggs	0.150	0.042
Grapefruit	0.075	0.021
Kale	0.205	0.058
Lettuce	0.088	0.025
Milk	0.053	0.015
Oatmeal, dry	0.55	0.156
Oranges	0.110	0.031
Peas, dry	0.46	0.13
—— fresh, green	0.38	0.108
Potatoes	0.130	0.036
Tomatoes	0.093	0.026
Walnuts	0.45	0.127

[a] Each figure is stated to the presumably significant number of places.
[b] These data are from the *Journal of the American Medical Association*, April 4, 1942.

liams originally spoke in terms of "restoration" of the thiamine content of flour to that of wheat, governmental administrative officers have preferred to think in terms of making the white bread "carry its share" of the thiamine that is needed in human nutrition. The standards set for "enriched" [2] flour and bread

[2] This name was decided upon by the Federal Food and Drug Administration.

by the Federal Food and Drug Administration are such as to bring the thiamine content to about the lower margin of the range of whole wheat.

Bread plays such a large part in the dietaries of the low-income population that, using food-consumption data of 1935–39 as a baseline, Williams estimated that if all white bread were enriched, this would increase the thiamine intake of the people of the United States by more than 60 percent. By 1942 the greater part was enriched and the nutritional status of our people thereby materially improved as regards thiamine. At the same time fuller employment with improved purchasing power has permitted increased use of foods which tend to give the diet better all-round mineral content and vitamin values, thus further raising the intake of thiamine along with other important nutrients. Hence it should no longer be regarded as unduly difficult to furnish thiamine in liberal amounts through the everyday dietaries of the people generally.

As an example which could be varied at pleasure: if one's breakfast includes 8 ounces of orange (or juice), 1 ounce (dry weight) of oatmeal or shredded wheat, and 1 glass (8 ounces) of milk; luncheon, 1 egg (2 ounces), 4 ounces of potato, 7 ounces of lettuce-tomato salad, 2 ounces of bread, a glass of milk, and 4 ounces of banana or grapefruit; dinner, 2 ounces of lean meat, 3 ounces of green peas, 4 ounces of carrots or potato, 3 ounces of cabbage or tomato, 2 ounces of bread, and a glass of milk; and at such times as one chooses an ounce of walnuts (or peanuts or peanut butter)—even if only half of the bread is whole wheat and even if the white half of the bread were not enriched, one would still receive from these foods about 2 milligrams of thiamine, while enrichment of the white bread would add about 5 percent to the thiamine intake. If, on the other hand, 6 ounces of bread were eaten daily, and all of it white, its enrichment would increase the day's thiamine intake by an amount equivalent to about 15 percent of that contained in a day's food such as just suggested.

The Pellagra Problem in the United States and Niacin
(Nicotinic Acid) as a Factor in Its Solution

Among the afflictions of southern Europe in the period of great poverty following the Napoleonic wars was a disease characterized by rough red skin from which it took the name *pellagra.*

About 1907 this disease was first reported as occurring in our South, especially in the villages and towns which had grown up rapidly, with little planning or provision for sanitation, around the textile mills then recently moved into the region largely because of the cheap labor supply. Here many families, always accustomed to low money income but previously living on independent subsistence farms, had become wage earners. As such they handled more money than they had before; but not enough to enable them to meet their increased cost of living and to provide good family dietaries, now that they were obliged to buy all the food they consumed. It was a rude awakening from the American dream of a New World affording good lives for all its people to realize the growing prevalence here of a disease previously identified with the destitution of a war-impoverished Old World.

Many of the mill towns reporting pellagra epidemics lacked adequate sanitation. Also, the victims of the disease were found chiefly in families recently removed from living on the land and unaccustomed to congestion in their housing.

Epidemiological studies of pellagra around 1910 led to two theories: one, that it was due to infection through some unknown organism; the other that it was a nutritional deficiency disease. The latter theory was supported by a striking experiment performed with human subjects by Dr. Joseph Goldberger of the United States Public Health Service.

The Governor of a Southern state decided to use his pardoning power for the promotion of the public health. He offered pardons to a dozen convicts (selected, of course, with due reference to both moral and physical qualifications) who volunteered to live for a year, if they remained in good health, upon what Goldberger believed to be a pellagra-producing diet. One of the convicts after living on this experimental diet for a time decided to return to the regular prison-farm fare. The others continued and well within a year several of them developed pellagra symptoms; all were pardoned and immediately departed, not waiting to be cured, as no one concerned in the experiment doubted that they would promptly recover on a good diet.

For convenience in remembering the chief features of the typical clinical picture, pellagra has sometimes been called the disease of the three D's: dermatitis, diarrhea, and depression. It used also to be said that any two of these three signs would justify a diagnosis of pellagra. If the dermatitis were lacking, it was *pellagra sine pellagra.* Sore mouth was frequent; and the disease known as blacktongue in dogs was the nearest analogue of pellagra to be experimentally producible and controllable under laboratory conditions. Late in 1937 it was discovered by Elvehjem of the University of Wisconsin that blacktongue could be cured by the substance long known to chemistry by the inappropriate name of nicotinic acid. At once this was tried upon pellagrins by at least four groups of medical research workers, and with highly gratifying results. So striking was the improvement of typical pellagra patients under nicotinic acid (niacin) treatment that pellagra soon came to be considered and sometimes formally defined as the niacin deficiency disease (*aniacinosis*).

The situation of most pellagrins, however, is not so simple. While niacin often cures in dramatic fashion the symptoms of pellagra, yet it does not usually restore the typical pellagra patient to full health. Often he also needs thiamin or riboflavin or both. Medical opinion is now well agreed that the poor pellagrin as actually met in practice is usually suffering from deficiency of more than one vitamin. Moreover, there are still two ways of expressing this view. If pellagra is defined as niacin deficiency, then the cases met in practice are rarely simple pellagra; while if pellagra means the condition of the typical pellagrin as actually met, then niacin is the most prominent factor in, but not a complete solution of, the pellagra problem. From this viewpoint blacktongue is not as closely parallel to pellagra as has hitherto appeared to be the case. Partly for these reasons and partly because analytical determinations of niacin in foods are not uniformly satisfactory or trustworthy, no table is here offered of either nicotinic acid contents or blacktongue-preventing values of foods.

It is more to the point of our present purpose to record the finding, from direct human experience, that pellagra can be prevented by a daily allowance per person of: a quart of milk or buttermilk; a pint of evaporated milk; one third to one half pound of lean meat or canned salmon; one pound of green peas, or kale, or fresh or canned collards or turnip greens; or half each of any two of these items; or the equivalent distributed as desired.

Doubtless the chief explanation is that these pellagra-preventive foods are good sources both of niacin and of riboflavin, and at least fair sources of thiamine as well, so that they make good the supplies of all three of these vitamins.

Riboflavin (Vitamin B₂ or G)

One reason for the chemists' lateness in discovering vitamins is that they occur in such small amounts as to have been missed in even very painstaking food analyses. Riboflavin, however, was a partial exception. In 1879, Blyth in reporting the results of his research analysis of milk, noted the presence of a very small amount of a water-soluble greenish-yellow coloring matter which he called *lactochrome*. Its color had resulted in its detection, and later when it was found to belong to the chemical group of natural colors known as *flavins*, its name was changed to *lactoflavin*.

During a decade, 1926–36, by a process of typically gradual discovery to which several laboratories contributed independently, it was found that this substance performs a part of the nutritional functions which had been attributed to vitamin B. In other words, the original concept of vitamin B was differentiated, first into vitamin B_1 (now called thiamine) and vitamin B_2. Then the latter was further differentiated into a number of "vitamins of the B_2 group," one of which was this substance, lactoflavin. Examination of other foods revealed the presence of either the same or closely analogous substances which tentatively were designated as flavins, with prefixes to indicate the natural source from which the material was obtained in each case. The new knowledge of their nutritional significance greatly stimulated research upon the physical isolation and complete chemical identification of these substances, and different groups of investigators then soon reached the conclusion that lactoflavin from milk, ovoflavin from eggwhite, hepatoflavin from liver, and presumably the nutritionally active flavins of other foods, are the same substance, which is now also made synthetically. With its molecular structure thus established, the full chemical name indicative of this structure was duly assigned but was much too cumbersome for daily use, so the name *riboflavin* was coined as indicating as much about the chemical nature of this substance as could be explained in one word of moderate length. The riboflavin contents of some typical foods are shown in Table 5.

In examining Table 5 as to riboflavin contents of typical foods, it is important (as already noted in connection with analogous tables) to keep in mind the differences of condition of different foods as to moistness or dryness. The widely different amounts and proportions in which different staple foods may reasonably be expected to enter into the normal dietary is also important. Thus while the riboflavin content of milk does not appear strikingly high in the table (milk being about seven eighths water), yet in ordinary normal dietaries, milk furnishes 35 to 60 percent of the total riboflavin of the food consumed.

Foods which experience and human experimentation had shown to have high pellagra-preventing value may be seen

TABLE 5

RIBOFLAVIN CONTENTS OF SOME TYPICAL FOODS
(EDIBLE PORTION)

Food	Milligrams per 100 g.[a]	Milligrams per ounce [a]
Apples	0.02	0.006
Bananas	0.05	0.014
Beef, lean	0.233	0.066
Bread	Status not yet determined	
Butter	trace	trace
Cabbage	0.051	0.014
Carrots	0.056	0.016
Cheese, Cheddar type	0.527	0.149
Eggs	0.371	0.105
Kale	0.50	0.14
Lettuce	0.06	0.017
Milk	0.207	0.059
Oatmeal, dry	0.123	0.035
Oranges	0.036	0.010
Peas, dried	0.225	0.064
—— fresh, green	0.187	0.053
Potatoes	0.038	0.011
Sweetpotatoes [b]	0.086	0.024
Tomatoes	0.044	0.012
Turnips	0.060	0.017

[a] Each figure is given to its presumably significant number of decimal places.
[b] Sweetpotato, as one word, is used by scientists to indicate the botanical differentiation from potato.

from Table 5 to be relatively rich in riboflavin. The "trouble" of the typical pellagrin as actually met is usually a multiple deficiency in which both niacin and riboflavin are important factors. If from now on the word pellagra is to be limited by definition to such part of the "clinical picture" as relates only to the niacin deficiency, then for the part attributable to riboflavin deficiency we have the old name, *pellagra sine pellagra,* and the new name *ariboflavinosis.*

It might also be said that if niacin is the outstanding factor in the disease problem of pellagra, riboflavin holds an equally outstanding position in the problem of the relations of nutrition to general resistance and to the constructive advancement of the norms and standards of human health. The American Medical Association, in extending official recognition to riboflavin, characterized it as essential to tissue resistance. Bessey and coworkers at the Harvard Medical School showed that the ability of the body to resist certain types of disease is very importantly influenced by the level of riboflavin intake. Undoubtedly too this also influences typical life histories in which specific diseases do not appear, higher levels of riboflavin in the habitual daily food tending to higher health and longer life, as will appear more fully in some of the subsequent chapters.

In a recent tentative survey at Columbia University with the aid of a grant from the Williams-Waterman research fund and other collaboration of Dr. R. R. Williams, we have estimated that of the total riboflavin of the American dietary of recent years 44 percent came from milk and cheese, 19 percent from meats including fish and poultry, 6 percent from eggs, 15 percent from breadstuffs and other cereals, and 16 percent from fruits and vegetables. It appears also from this preliminary survey that the per capita consumption of riboflavin lies within the zone between the level of minimum requirement and that needed for best nutrition, and still would lie in that zone if "the enrichment program" should result in doubling the amount furnished by breadstuffs.

These facts taken in conjunction with the additional experimental evidence of the relation of riboflavin to health, to be cited in later chapters, give this vitamin a place of high importance in the nutritional opportunities and responsibilities of the present and the future.

CHAPTER VII

FAT-SOLUBLE VITAMINS A AND D

Vitamin A and Its Far-reaching Relations to Health

THE existence of the substance which came to be called vitamin A was discovered in 1913 through experiments made independently by McCollum and Davis at Madison and by Osborne and Mendel at New Haven, in which it was found that young experimental animals on diets good in other respects would grow and develop or would stop growing, sicken, and die, according to the fat chosen for the dietary. This was found to be because some food fats do, and others do not, contain dissolved in them significant amounts of a nutritionally essential fat-soluble substance.

The existence of this substance having thus been discovered through growth experiments, the essential relation of vitamin A to the growth and development of the young was outstandingly emphasized in all of the earlier literature. However, it is now known that vitamin A is essential at all ages. The fact that the young are growing makes them sensitive reagents for experiments demonstrating that the substance is essential; but by properly arranged long-time experiments, it has more recently been shown that the elderly need much more vitamin A than was first supposed.

In one of its functions, vitamin A is concerned in vision: it takes an essential part in the chemical reactions upon which visual perception and the regeneration of the substances active in the visual process depend. Hence, one effect of a shortage of vitamin A is a diminution of ability to see in a dim light or to adapt to changes of intensity of light. This condition, long popularly known as night blindness, has recently come into prominence as a factor in the frequency of automobile accidents. An interesting instance of its industrial significance was observed in the fact that workmen receiving more vitamin A were more efficient in matching the whiteness of tiles.

In a second, more generalized, function, vitamin A is essential to the maintenance of the integrity and resistance of mucous membranes. Even in their early experiments, Osborne and Mendel observed that rats on diets deficient in the fat-soluble vitamin became subject to an eye disease which has been variously called an ophthalmia, or xerophthalmia, or conjunctivitis. A little later, the more detailed pathological investigations, especially of Wolbach of the Harvard Department of Pathology, indicated that the underlying cause of the conjunctival eye trouble and also of the increased incidence of respiratory disease is a microscopic breakdown of the cells of the mucous membrane and their displacement by a different form of tissue. More recently Kruse (1941) has studied in detail the phenomenon which, at least in certain stages of its development, has been known as Bitot spots—a roughness of the mucous membranes particularly around the angles of the eyes. By the use of a method of improved delicacy, he finds a very high incidence of this condition which, inasmuch as it is curable by vitamin A, he interprets as being a delicate criterion of slight degrees of vitamin A deficiency. Abnormal conditions of the skin also develop under conditions of shortage of vitamin A intake.

Whether the first sign of a vitamin A deficiency in the human body will be a diminished efficiency of adaptation of vision to a changed intensity of light (dysadaptation, night blindness, hemeralopia), or changes in epithelial cells of mucous membranes such as may lead to replacement of a different type of tissue or to the development of Bitot spots, or whether a less specific decline of condition in the skin will first appear, may depend in part upon individual differences in the persons affected and in part upon the acuity and interpretation of the observer. For any one of the three indications may seem more delicate or more dependable to one investigator than to another. Also it is readily conceivable that when all the signs are equally understood, one may be found more delicate for the detection of a chronic slight shortage, while another may permit of earlier detection when the shortage of vitamin A intake is more drastic in degree.

The physical and chemical properties of vitamin A are such that the body when receiving more than enough for its current needs can store a good proportion of the surplus, so that at a given time one individual may be carrying a sufficient bodily reserve of vitamin A to last him a relatively long time, while another person may have almost no such reserve, depending upon their respective backgrounds of food habit. The readiness with which the bodily stores of vitamin A (most of which is held in the liver) will be mobilized into the blood and distributed to all the tissues to meet a condition of lowered nutritional intake may also vary with circumstances, or from one individual to another. Thus the vitamin A content of the blood depends in part upon the amount currently received in the food and in part upon the bodily reserve stores. Measurements of visual adaptation have not always been found to run parallel either with the dietary history or with determinations of the vitamin A content of the blood.

At present, therefore, we know relatively little about the extent of the differences between individuals as to the amounts of vitamin A they need. Nor has there yet been time for the development of a full consensus of expert opinion as to the frequency of the bodily condition of vitamin A deficiency, or how best to diagnose it. Yet, viewed as a whole, the evidence now available makes it sufficiently clear that liberal intakes of vitamin A are important at all ages.

The Table of Recommended Daily Allowances, published early in 1941 by the Committee on Food and Nutrition of the National Research Council, and frequently called "the new yardstick of good nutrition," provides 1,500 International Units of vitamin A for children under one year of age; 2,000 from four to six; 3,500 from seven to nine; 4,500 from ten to twelve; and 5,000–6,000 for all people over twelve, with an increase to 8,000 for women during lactation. The trend of both clinical and laboratory research since the time that these allowances were proposed has strengthened the view that they are none too liberal. If they were now to be challenged at all it would probably be for their apparent implication that (except for nursing mothers) the individual's maximal vitamin A requirement has been reached at the age of twelve. This implication, however, may be construed as meaning not that children in their early teens necessarily need as much vitamin A as their elders, but rather that, with their good appetites and assimilative powers, theirs is a good age at which to establish food habits ensuring a liberal vitamin A intake. Also, liberal intakes early in life tend to a desirable sort of "insurance" in the form of bodily reserves of vitamin A, built up early and maintained throughout life.

For we now have good scientific reason to expect a higher level of positive health and a lessened loss of time through disease, when we live on liberal rather than minimal nutritional levels of vitamin A. The evidence of a distinct lowering of the death rates of the latter half of the life cycle is being investigated further.

The evidence that the incidence or severity of infectious disease or the duration of a minor infection such as the common cold may be influenced by the amount of vitamin A furnished by one's food has been considerably debated. Many physicians have felt it a matter of professional duty or ethics to use their influence in offsetting what they regarded as exaggerated hopes in this direction. Thus there has been a period in which medical literature has been especially skeptical toward the human importance of vitamin A and its bearings upon infection. Yet more recently this skepticism has been largely replaced by acceptance of the steadily growing weight of experimental evidence. The *Journal* of the American Medical Association has suggested that in the light of Wolbach's research, vitamin A must be accorded an important place in the maintenance of the integrity of the body's "first line of defense"—the normal epithelium of the mucous membranes. Certainly, too, vitamin A has some deeper influence as well, for in researches both by Lassen in Copenhagen and by Boynton and Bradford in the University of Rochester Medical School it was found that vitamin A in some way helped the body to cope with experimental infections even when these were administered by subcutaneous or intravenous injection. Similarly convincing were the experimental inoculations made by McClung and Winters at the University of Texas. Conciseness forbids multiplication of instances.

The case of the common cold with its related minor respiratory infections is of general interest to all. Here the reported findings are not uniform. Sometimes no clear-cut results have been found. In other investigations there has appeared to be an unquestionably lower incidence of such infections as the common cold among people receiving extra vitamin A than in control groups. In still another study no clear difference of incidence appeared, but the people having higher intakes of vitamin A recovered more quickly from infections of a given severity. The evidence as a whole therefore leaves us with the impression that vitamin A is by no means a sure preventive of colds or respiratory infections, but that liberal intake of this vitamin probably has value in reducing sometimes the incidence, sometimes the severity, or sometimes the duration of these infectious diseases.

It may be difficult to discriminate between adding a surplus and meeting an unrecognized deficiency. If the delicate diagnoses of Kruse (1941) are accepted as specific of shortage of vitamin A, then deficiencies are more frequent, and presumably requirements are higher, than had been supposed. In the simultaneous work of Yarbrough and Dann (1941), the level of vitamin A in the blood appeared to be the most promising single criterion. At the same time Leong (1941) found that while in general the vitamin A content of the blood tends to be proportional to the vitamin A value of the intake, a low level in the blood cannot be considered as conclusive evidence of depletion of the bodily reserves.

In the April, 1942, *Journal of Nutrition*, Lewis, Bodansky, Falk, and McGuire report the results of a very interesting experimental investigation of the relation of the level of nutritional intake of vitamin A to growth and to the concentration of this vitamin in the blood plasma, the liver, and the retina. Starting with rats three to four weeks of age (which may be considered as "end of infancy"), they found that feedings of two International Units of vitamin A per young rat per day prevented gross and histological evidences of vitamin A deficiency, and most of these animals had maximal vitamin A concentrations in the retina. The investigators therefore concluded that this intake may be regarded as fully adequate for immediate requirements; but they found it insufficient to support any appreciable vitamin A reserve in the liver or as high a level in the blood plasma or as high a growth rate as did more liberal levels of feeding. As judged by gain in weight, the optimal intake of vitamin A was between 5 and 12 times the minimum requirement as above defined, and only when the intake was again doubled was the plasma vitamin A maintained at the apparently optimal level of 100 units per 100 ml.

In this same series of experiments these investigators found that an intake (per young rat per day) of 25 units supported only a negligible reserve in the liver; 50 units, a reserve of 34 units per gram of liver; while an intake of 100 units resulted in an average liver reserve of 113 units per gram, which enabled the animals to maintain good health during a subsequent period of deprivation of vitamin A for 20 weeks (about one seventh of the natural life cycle).

In the Columbia Department of Chemistry, researches upon the influence of different degrees of liberality in the vitamin A value of the food have been carried on for several years. The early experiments of Dr. F. L. MacLeod demonstrated clearly that liberal intake of vitamin A is important for adults as well as for growing children. Those of Dr. E. L. Batchelder showed successive improvements, in the life histories of experimental animal families through two generations, by means of step-wise enrichments of the diet in vitamin A up to about four times the level of minimal adequacy. In those of Dr. H. L. Campbell and her coworkers, it has been found that food supplies of higher vitamin A value lower the death rates among adults even more markedly than among the young. Noteworthy also is evidence that, in the young, the body can store vitamin A upon a diet of a given composition and that later in life a diet of the same composition proves insufficient to meet the current rate of expenditure, so that the older individual shows diminution of bodily store of vitamin A.

Table 6 shows the vitamin A values of several types of foods.

TABLE 6

VITAMIN A VALUES OF SOME TYPICAL FOODS
(EDIBLE PORTION)

Food	International Units per 100 g.	International Units per ounce
Apples	70	20
Bananas	250	70
Beef, lean	30	8
Bread	traces	traces
Butter	5,000	1,400
Carrots	3,000	850
Cheese, Cheddar type	2,000	570
Eggs	1,500	425
Egg yolk	4,000	1,130
Greens (kale, spinach, etc.)	20,000	5,700
Lettuce, headed	around 300	around 85
—— loose leaved	" 3,000	" 850
Milk	200	57
Oatmeal, dry	traces	traces
Oranges	200	57
Peas, fresh, green	1,200	340
Potatoes	40	11
Sugars and Sweets	none or negligible	
Sweetpotatoes	2,500	700
Tomatoes	850	240

In speaking of foods generally, or of food supplies or dietaries, we say vitamin A "value" rather than "content." This is because foods may not only contain vitamin A itself but also (or instead) may have vitamin A value because of containing substances which change into the vitamin in our bodies, or in the bodies of other animals. The chief of these *precursor* substances are the carotenes, taking their name from the carrot to which they give its characteristic orange-yellow color. Green leaf vegetables are also relatively rich in carotene, which, however, is hidden (to the unaided eye) by the deep green color of the chlorophyll. When for the practical purpose of dietetics and the planning of food supplies, foods are classified into some ten to twelve groups, of which one is, "Leafy, green, and yellow vegetables," the reason for grouping these vegetables together is the high vitamin A value which most of them share. The few exceptions are chiefly due to the fact that some yellow vegetables owe their color to substances other than the carotenes. Moreover, the carotenes differ among themselves in vitamin A value, beta-carotene yielding twice as much vitamin A as does an equal weight of alpha-carotene. Hence our estimates of vitamin A values of foods are more trustworthy when based upon well-conducted feeding experiments with the foods in question, rather than upon measurements of carotene content.

A Very Short Story of the Vitamins D and the Eradication of Rickets

Whether it is an exaggeration to speak of the eradication of rickets may depend upon where one lives. To many regions in the temperate zones, at any rate, the newer knowledge of nutrition has brought among its other benefits such better-built bodies as make today's a more sightly humanity than that of previous generations and one freer from such faults in the pelvic bones as hitherto handicapped many women in childbearing. Rarely now do we see the bowed legs, the enlarged or knobby joints, or the misshapen chests which were so common for centuries until well within our own. The food shortages of Central Europe during the First World War brought an increased incidence of rickets and at the same time that this became clearly known to our English-speaking world the means of preventing this disease began to be clarified.

Mellanby in England found that with other conditions the same, he could cause or prevent rickets in puppies according to the nature of the fat that he put in their food mixture. Cod-liver oil was very potent, and butterfat was moderately effective, in the prevention of rickets; and McCollum, in experiments at the Johns Hopkins School of Public Health, showed that this rickets-preventing potency was not lost when the vitamin A in the fat was destroyed by heating with exposure to air.

Vitamin D was the term adopted for the fat-soluble rickets-preventing substance thus clearly differentiated. At about the same time it was rediscovered that adequate sunshine also prevents rickets, and, further, that this effect can be referred to the ultraviolet rays of a certain range of wave length.

Thus for a few years it was a matter of frequent remark in medical discussion that while for (or rather against) many diseases the physician had no "specific," against rickets he had two: on the one hand, ultraviolet light; on the other, cod-liver oil and, in lesser degree, the oil of egg yolk and the fat of milk.

Then in 1924 Hess in New York and Steenbock in Wisconsin independently discovered that other foods could be endowed with the antirachitic property by ultraviolet irradiation. Rapidly then it became apparent that "the two specifics" against rickets were essentially two ways of introducing "vitamin D" into the system: it can be swallowed as such, or it can be produced in the skin by the action of ultraviolet light upon one or more of the *sterol* substances which normal skin always contains.

Here we note also an interesting provision of nature. The same bright sunshine (or artificial light containing an adequate proportion of ultraviolet) which produces vitamin D in the skin also tends to sunburn which involves as an early or preliminary stage a flushing due to increased circulation of blood through the exposed skin. This increased rapidity of the local circulation carries the new-formed vitamin promptly into the interior of the body where it is safe from overexposure to the ultraviolet, and where it enters at once upon its functions or, in times of abundance, may be stored for future use. Industrial production of vitamin D by irradiation has not always been so effectively safeguarded against overexposure, which may cause formation of undesirable by-products. This is one reason for the preference sometimes expressed for natural rather than artificially produced vitamin D; but another reason is that "vitamin D" really stands for a group of antirachitic substances (the vitamins D) of which one predominates in natural, and another one in artificially produced, antirachitics. This difference between natural and synthetic vitamin D affects rats very little, but chicks a great deal. How much difference it makes to children, the pediatricians do not yet agree.

Also there is still considerable doubt as to just how vitamin D performs its function or functions in the body of the baby. Undoubtedly it aids assimilation of calcium and phosphorus in cases where assimilation is otherwise poor. But whether this function extends beyond the mere facilitation of absorption of either or both of these elements from the digestive tract is still debated. There are strong reasons to believe that, however its effects may ultimately be explained, vitamin D may not only prevent rickets but also serve constructively to aid the growth of the baby in length of body and thus ensure realization of the child's full birthright as to stature.

It is doubtless also true that individuals who during infancy have been protected from rickets by cod-liver oil, or any of the other liver oils, are less subject to respiratory diseases both at the time and afterward. For this, there well may be two good reasons: (1) the better-formed and therefore better-functioning thorax built once-for-all in the individual who had plenty of the right kinds of vitamins in infancy and early childhood; and (2) the continuing protective value of the fuller bodily reserve of vitamin A which is liberally supplied along with the vitamin D of the natural liver oils and, in lesser degree, by the fats of eggs and of milk.

There are, of course, other fat-soluble vitamins, notably the vitamins E and the vitamins K. It seems better, however, not to give space to them here as their significance is so largely medical as to lie rather outside the scope of this book.

CHAPTER VIII

HOW THE BODY MANAGES ITS NUTRI-
TIONAL RESOURCES

T HE body is much more than a machine. True, there has been a fashion of speaking of bodily systems as mechanisms, and such a way of speaking was doubtless helpful in the days when science was learning that the principle of the conservation of energy holds good in our bodies very much as in our mechanical inventions. Yet the body does so much more for itself than any machine can do that to speak of bodily processes as mechanisms would be very misleading if we did not consciously or unconsciously carry in mind that in science the suffix *ism* is usually to be taken as signifying overemphasis of a particular point of view. Thus the distinguished biochemist William Mansfield Clarke in a discussion of scientific theory pointed out that "-*isms*" become "*wasms*." Without assuming to speak for all of biology, it is appropriate here to note that in the science of nutrition where biology meets and is interwoven with the more exact sciences, "mechanism" as a supposed characterization and explanation of our bodily systems and processes is now rapidly becoming a "wasm." True enough, the total energy manifestations of the body can be correlated with, and in the long run are dependent upon, the nutritional fuel which the food as a whole supplies. True enough also, some parts of bodily structure, and some aspects of their working, can be illustrated and in part explained by machine model analogies. Yet it is still more significantly true that the body does so much more for itself than any machine model can do that mechanical analogies are but partial and inadequate.

Thus the body may, if one wishes, be regarded as somewhat

machinelike in the movements of its muscles and joints and in the long-run quantitative relations of its intake of fuel and output of kinetic energy; but the chemical reactions upon which the energy transformations depend are in the body speeded-up by catalysts which the body itself makes. Here we have a kind and degree of self-acceleration quite beyond anything that machine models have or can adequately represent.

Catalysts such as we have just mentioned, formed in living cells from the organic substances which they contain, are called *enzymes*. It is by virtue of the speeding-up effects of enzymes as catalysts that the body's chemical reactions are made to run fast enough to support its life processes. This is true both of the processes in the digestive tract and in the body tissues; and in the latter it is true both of anabolic (building up or "assimilation") and catabolic (breaking down or "degradation") processes. What one speaks of as *a* nutritional process may involve a whole chain of chemical reactions. It need not be assumed that all the reactions of such a chain or series are equally dependent upon catalysts. Typically, however, one or more of the chemical reactions essential to a nutritional process is enzymic, that is, needs the catalyzing effect of some enzyme "to make the process go fast enough to keep our lives going." In other words, all the higher forms of life are highly catalyzed as well as chemically complex affairs.

Our *digestive enzymes* include several distinct substances and all of these which have been adequately investigated have been found to be, or to contain, proteins in their material make-up. Obviously their enzymic properties (which, it should be remembered, have somehow been built into them by the body itself) make them something more than ordinary proteins; yet nearly all of the material substance of each of them can be accounted for in terms of the same amino acids which we absorb as digestion products of our ordinary food proteins.

This is one illustration of the "processing" or "chemical

engineering" of ordinary food materials into highly specific body substances which we have sought to suggest by use of the phrase "how the body manages" in the title of this chapter.

What has just been said about the chemical natures of the digestive enzymes is true also of *insulin,* the active substance of the internal secretion of the pancreas which catalyzes one or more phases or steps in the metabolism of carbohydrate.

Glutathione, a constituent of active tissues generally and a catalyst of some of their nutritional reactions, is a much simpler substance, containing in its molecule only three amino-acid radicles as against a great many in a molecule of insulin or of any typical digestive enzyme. Yet here again it is true that the body has made its specific catalyst from the amino-acid digestion products of its ordinary food proteins. And this is also true of both the best-known catalysts which the tissues receive from other sources: *thyroxine,* from the thyroid gland, and *adrenine* or *epinephrine,* from the adrenal glands, is each a derivative of an amino acid such as the body derives from its protein food.

Thus the body so manages the digestion products of its food-protein supply as to make them serve both the "building block" function of general tissue growth or upkeep and the more special functions of precursors from which the body makes the enzymes it needs to keep its nutritional processes going fast enough.

Many man-years of work were consumed in the task of ascertaining the chemical natures, and thus the nutritional antecedents, of these essential enzymes which the body makes for itself from materials supplied by the food. Parallel enzymes are found in plants and function in their important processes from the first beginnings of the sprouting of the seed. One of these plant enzymes was studied by the same Dr. Osborne who contributed so much to the knowledge of the chemical natures of food proteins. Osborne purified malt amylase and reported it to be of protein nature. Researches at Columbia confirmed his conclusion and extended the study to the investigation of pancreatic amylase, the enzyme formed in the pancreas and functioning in our digestion of starchy food.

Employing several different combinations and modifications of the processes of purification, workers in the Columbia laboratory repeatedly obtained preparations of pancreatic amylase of apparently uniform chemical and physical properties and of practically constant maximum enzymic activity; and this product, finally obtained in crystalline form, was, like the malt amylase previously prepared by Osborne, of protein nature.

In 30 minutes at body temperature, this material digests about 20,000 times its weight of starch and forms about 10,000 times its weight of maltose. Although subject to serious deterioration in solution, the activity of this enzyme preparation was such that when tested in longer experiments it digested 4,000,000 times its weight of starch and formed no less than 2,800,000 times its weight of maltose before it had all become inactivated.

This marked enzymic activity was exhibited by the preparation at a dilution of 1:100,000,000 parts of water. The most delicate tests for proteins are not valid at dilutions greater than about 1:100,000. The material when tested in concentrated form reacted like typical protein to the usual protein tests, but its own enzymic activity constituted a test for its presence which was 1,000 times (a thousandfold) more delicate. Thus the failure of protein reactions in solutions enzymically active did not show that the enzyme is of other than protein nature in its chemical composition, although this negative conclusion was erroneously drawn by some investigators and repeated by many writers.

Here it is of interest to recall the experience of the Curies in the concentration of radium from pitchblende. It was only after they had intensified the property of radioactivity fiftyfold by concentration of the radioactive element that its presence was revealed analytically even by so delicate a test as the use of the spectroscope. Delicate as is the spectroscopic test, the property of radioactivity constituted a test at least fifty times more delicate.

The brilliant work of Dr. J. H. Northrop proved the protein nature of several typical digestive enzymes with entire conclusiveness. Northrop (1939) stated the case unqualifiedly as follows: "The crystalline enzymes which have been prepared are proteins, but their activity may easily be detected in solutions which are too dilute to give positive protein tests." And it was with relation both to digestive enzymes and other body enzymes also that he wrote: "One of the most striking characteristics of living things is the rapidity and precision with which the chemical changes necessary for their existence are carried on."

Largely by virtue of the enzymes which abound not only in its digestive and internal secretions but within its active tissue-cells as well, the body holds its nutritional assets (or a considerable fraction of each of them) in a highly dynamic state of mobility. While some profess to find the term paradoxical, it is matter-of-fact knowledge today that the so-called "steady states" (or *homeostases*) in the body are quite as truly states of an *actively dynamic equilibrium*. This has been increasingly recognized for at least two decades past but has been specifically signalized in the recent publication, by the Harvard University Press, of a book entitled *The Dynamic State of Body Constituents*. In this book, begun by the late Dr. Schoenheimer and finished by Professor Hans Clarke, the keynote is the statement that all constituents of living matter "are in a steady state of rapid flux."

This is not as revolutionary as it may sound to some readers. In the first and second decades of this century the general dynamic physico-chemical view of living matter won in the minds of some students and teachers of nutrition a dominant place over the view which rested on a combination of the "machine-model" or architectural, and relatively static structural concepts of conventional organic chemistry. The more dynamic view was reflected in a period of active experimental study of *autolysis* in various organs and tissues removed with as little alteration as possible from healthy living bodies and then left to themselves at body temperature, but protected from contamination by microörganisms. The study of autolysis furnished much evidence (1) that the bodily organs and tissues normally contain enzymes of their own, analogous to the digestive enzymes, and (2) that the normal state of any typical living tissue is better represented by a concept of dynamic equilibrium (or near-equilibrium)—between, for instance, amino acids and tissue proteins—than by a concept of a structure in the machine-model or rigidly architectural and building-block sense. In recent decades the dynamic-equilibrium view has been active in the continuous systematic studies of body enzymes while the more mechanically structural concept is still reflected in such expressions as "wear-and-tear" of tissues and "repair" functions of nutrients derived by the body from the food.

Especially important to our present discussion is the new evidence obtained through experiments with isotopes.[1] This tends strongly to the more dynamic mode of nutritional thought.

As Schoenheimer put it: "Food and tissues both consist of proteins, fats, and carbohydrates. When the split products liberated by digestion are absorbed, they merge with identical molecules which originate from the tissues"; to the extent that these molecules are strictly identical, the investigator may himself lose track of which is which and certainly cannot attribute to a chemical reaction proceeding in a body

---

[1] While the isotope is identical in chemical behavior with the familiar form of the element, it is "tagged" or "labeled" by having a heavier or lighter atomic weight, so that solutions of its compounds have corresponding differences in physical properties. Thus deuterium ("heavy hydrogen") forms heavy water. Except for deuterium the isotopes are not given separate names but are distinguished by giving the atomic weight as a superscript number with the atomic symbol, e. g., N^{15} for heavy nitrogen.

Radioactive isotopes formed artificially by electron bombardment of elements are sometimes indicated by an asterisk; e. g., radioactive phosphorus may be written P^*. In the use of these isotopes the radioactivity is a quick means of tracing the atoms thus tagged.

cell any power to distinguish between, say, the amino-acid tyrosine absorbed from the digestive tract and tyrosine liberated from tissue protein. Some earlier investigators did find means of following the fates of nutrients in the body farther and more definitely than the literature of isotope experimentation might lead one to suppose; but the use of isotopes as tagged atoms of nutrient substances gives a new order of elegance and convincingness to such study.

Since the discoveries of Urey, chemists can synthesize organic nutrients containing isotopic elements which have no effect upon the fate of the "tagged" nutrient molecules or radicles in the body and do permit the experimenter to keep track of what he introduced through feeding in contradistinction to what the body already had. Of course, the "keeping track" is in a sense statistical rather than a following of individual atoms or molecules; but as this is true of chemical investigations generally, the investigator's whole background of study and research experience in chemistry has inured him to this degree of handicap.

In what follows, the body fats are considered first and then the body proteins.

Most of the fat in our bodies, as in those of other mammals, is held in the so-called fat depots: between the skin and muscles, and around the internal organs. This tendency to localization, together with the relative insolubility of fats in water and most watery fluids, has generally been taken as meaning that stored fat remains chemically inactive unless and until an energy deficit in the body's nutritional account creates a need for bringing some of this reserve fat into the circulation for use as fuel. Now, however, it is found that even while there is no energy deficit and the body's store of fat remains essentially constant in amount, there is a relatively free interchange between the isotopically tagged fatty-acid radicles entering the body as digested food and the untagged but otherwise identical fatty acids of the body's depot fats. Thus the fat stored in the body is shown to be in a fairly mobile

state of dynamic equilibrium with the body's circulating resources, undergoing a very real exchange of material regardless of the question of surplus or deficit. In a typical experiment, rats were fed a diet containing 6 percent of butter
(selected as a food fat which contains "all the fatty acids which
the animals require"), to which was added about one tenth
as much of an experimental fat in the form of isotopic palmitic
acid, which naturally merged during absorption with the
ordinary palmitic acid of the butter. After 8 days of such
feeding, the body-fat was found to contain almost half of the
tagged fatty acid that had been fed. Giving all due weight
to the fact that human beings do not run through their life
processes as rapidly as do rats, these experiments must still
leave us with a more mobile picture than previously conceived of the body's active management of even its relatively
insoluble and segregated nutrient resources. What had previously appeared to be merely passive reserves are now seen
to belong to the general pool of actively working capital to a
larger and freer degree than hitherto conceived.

Of still more vital interest are the dynamic relationships
in the field of protein, amino-acid, and ammonia metabolism.
Here nitrogen is the key element. The healthy, full-grown
body tends to eliminate as much nitrogen as it receives from
the food. That is, it tends to maintain nitrogen equilibrium,
which normally means protein equilibrium also. Thus the
total amount of protein in such a body tends to remain nearly
constant, and presumably there is still greater constancy in
the composition of the body protein as a whole. Notwithstanding this, the body proteins and their amino acids exist
in a highly dynamic state.

Heavy nitrogen (N^{15}), which like heavy hydrogen was made
available by Urey's researches, has been incorporated in amino
acids otherwise the same as those which the body receives as
digestion products of food proteins, and these isotopic amino
acids have been fed under controlled conditions of experiment.

According to an earlier concept, when such isotopic amino

acids were fed to full grown animals in nitrogen (protein) equilibrium, much the larger part of the heavy nitrogen of the extra amino acid fed should have appeared promptly in the urine; but this did not happen. In typical experiments with two amino acids less than half of one and less than one third of the other was thus excreted. The body proteins had acquired the largest part of the tagged amino acids, evidently because these proteins are constantly undergoing exchange of material with the amino acids entering the body as digestion products of the food. Not all bodily organs are equally active in the fixation of dietary nitrogen or amino acid. Relatively more is obtained by the internal organs and the serum proteins, relatively less by the muscle proteins, and still less by the proteins of the skin. In absolute amounts, the muscle proteins get the most because their total mass is so great.

A very interesting fact observed by Waelsch and Rittenberg (1941) is that glutathione, the tissue catalyst above mentioned, assimilates its quota of dietary amino acid much faster than do body proteins.

The heavy nitrogen which the body gets when an isotopic amino acid is fed is not found exclusively in the form of that kind of amino acid in the tissue proteins. Amino acids undergo such interchange with each other that when one isotopic amino acid is fed, others are (in this sense) formed in the body from it.

Moreover, certain substances formed from other foodstuffs in the body may react with ammonia to form amino acids (which are in constant interchange with protein), and so can bring into the service of the body proteins even such an inorganic form of nitrogen as ammonia. In fact, ammonia salts (and correspondingly urea) may have protein-sparing value both in this way and in another—through the influence of microörganisms in the digestive tract which may actually synthesize protein from ammonia and suitable organic matter as yeasts and crop plants do. Experimental evidence thus requires that we recognize in the phenomena of protein-sparing by the other organic foodstuffs much more than the term in itself might imply.

The radioactive isotope of phosphorus has also been much used for the investigation of nutritional processes; and here again the findings show the body to have a greater flexibility of management of its nutritional resources than previously supposed.

Iron isotope has also been employed in the study of nutritional problems of the blood by several investigators at the University of Rochester; and other isotopes have been employed in studying the permeabilities of body membranes and in many other ways.

Several of the vitamins are now known to function in the body as essential parts of its tissue enzymes. The human organism, receiving thiamin, niacin, and riboflavin through its digestive tract, has power within its tissues to use these vitamins (and perhaps others) in making the enzymes that make the nutritional processes go fast enough for our needs. This dynamic concept is important not only philosophically or from the viewpoint of pure science, but also in our attitude toward such practical questions as arise in the problems of wartime food rationing and how to plan the best use of food resources in postwar nutritional rehabilitation; because the more dynamic view gives us more confidence in the flexibility of nutritional processes, and this should enable us to work with greater scientific freedom in the utilization of all available food supplies.

CHAPTER IX

NUTRITIONAL CHARACTERISTICS OF THE CHIEF GROUPS OF FOODS

CHAPTERS I and VIII have dealt with general aspects of the nutritional process, while in Chapters II to VII we have reviewed the more outstanding of the specific nutrients or nutritional factors, and in several cases have listed a number of typical foods to show how they compare quantitatively in these respective aspects of nutritive value.

In the present chapter, we address our attention directly and predominantly to the foods themselves, thus linking our preceding consideration of nutritional needs and processes with the questions of best use of food resources which are to be discussed in later chapters.

Food commodities can be grouped in different ways according to one's viewpoint. The viewpoint of the present chapter is nutritive values, or nutritional characteristics. Even from this one point of view, several different groupings are more or less current, not so much because of differences of opinion on the facts, but rather in accordance with the degree of detail that best suits the purpose in hand. What would be commendable discrimination in one discussion would be undue particularization in another. For purposes such as the one on which we are here engaged, a simple grouping of food commodities into half-a-dozen to a dozen categories is usually deemed sufficient.

Thus if we start by dividing all our familiar articles of food (food commodities) into a very few groups according to their outstanding nutritive values, the result may be somewhat as follows:

(1) Breadstuffs and other products of the grain crops: eco-

nomical as sources of energy and protein, but not at all well balanced (for the purposes of our nutrition) in the mineral elements and vitamin values.

(2) Sugars and fats: chiefly significant, in the nutritional sense, as supplemental fuel foods, though some fats are also important as sources of fat-soluble vitamins.

(3) Meats, including poultry, fish, and shellfish: rich in protein, or fat, or both, but having in general about the same calcium and vitamin shortages as do the grains, except that lean meats contain relatively more riboflavin and niacin (nicotinic acid).[1]

(4) Fruits and vegetables: highly important as sources of mineral elements and vitamins, but differing rather widely among themselves as to richness in particular elements and individual vitamins, as well as in protein and energy values.

(5) Milk: important as source of energy, protein, mineral elements, and vitamins; the most effective of all foods in ensuring a well-balanced dietary.

(6) Eggs: in a general way intermediate in nutritional character between meat and milk.

The foregoing omits nuts, or brackets them as an afterthought with the dry legumes—confessedly not much like the more succulent vegetables and the typical fruits. Cheese, cream, and ice cream may be considered, from a general nutritional viewpoint, as forms of milk.

Already it will be apparent to the reader that to divide all foods into only six groups will mean that a given group may contain articles of food which differ rather considerably from each other in some nutritional characteristics, and perhaps may appear still more widely different from the viewpoint of meal planning.

[1] In this book we use the word meat as meaning what it does in our general food supply, about 96 to 97 percent of muscle meat (with more or less of adipose tissue) and about 3 to 4 percent of liver. There is no way of making liver a larger part of the meat supply, though the liver may be diverted more largely from the general meat supply to those individuals who for medical reasons are judged to need it.

For the combined purposes of (*a*) the planning of conventionally acceptable meals to fit different conditions of food supply and purchasing power, and (*b*) the planning of production programs with due reference to the nutritional needs of the population, the United States Department of Agriculture makes use of a classification of foods into twelve groups as follows: (1) milk; (2) potatoes and sweetpotatoes; (3) dry mature beans, peas, lentils, and nuts; (4) tomatoes and citrus fruits; (5) leafy, green, and yellow vegetables; (6) other vegetables and fruits; (7) eggs; (8) lean meat, poultry, fish, and shellfish; (9) breadstuffs and cereals; (10) butter; (11) other fats; (12) sugars. Here the sequence is largely one of convenience of statement and partly also convenience of agricultural and household planning.

What is more to the immediate purposes of this book is to note in what ways and why the sixfold grouping first cited is subdivided in course of being expanded, as in the second list, from six to twelve. Starting then in the sequence of the earlier and simpler grouping, we find no subdivision of the group comprising the breadstuffs and other grain products. Sugars, however, are separated from fats, and fats are subdivided into two categories: butter; and other fats. Meats, together with poultry, fish, and shellfish, remain an unchanged group except to the somewhat variable extent to which those meats which are mainly fat may be grouped with "other fats" rather than in the meat bracket. From the original fruit-and-vegetable group, the twelvefold grouping sets off: tomatoes and citrus fruits (sometimes with the addition of raw cabbage) as an independent group justified by the importance of their richness in vitamin C; and the leafy, green, and yellow vegetables on similar consideration of vitamin A value. Milk (with those of its products which sufficiently retain its nutritional character) and eggs each stands in the same position in either the sixfold or the twelvefold grouping.

At a relatively early stage in the development of the newer knowledge of nutrition, McCollum had found that many

American dietaries consisted so largely of bread, potatoes, meats, and sweets as to be in danger of shortage of calcium, or vitamin A value, or both. He then proposed the term "protective" for the two types of food (milk and green leafy vegetables) which are rich both in calcium and in vitamin A value. A little later, with growing appreciation of vitamin C, it became customary to include fruits, vegetables, and milk (including cheese, cream, and ice cream), with or without eggs, under the term "protective foods." Eggs were sometimes included and sometimes not. For the egg is relatively rich in vitamin A with a fair quota of calcium and fair to good proportions of most other nutrients, but it lacks vitamin C and also it does not help to maintain a good intestinal hygiene or to support the body's alkaline reserve as do fruits, vegetables, and milk.

The enrichment of white flour and bread with thiamin, niacin, and iron is, as explained in Chapter VI, a very important improvement; but it is only a partial restoration of what the whole grain contained and certainly does not lift bread out of the clear and useful classification of "breadstuffs and other grain products." Similarly the fact that determinations in one laboratory indicate moderately higher average B-vitamin values for meats than was found in other laboratories before [2] and since,[3] does not call for any change of the familiar position of meats in a general grouping of foods.

Diets containing preponderant proportions of bread and meat are still seen to need nutritional balancing in the way indicated by McCollum when he introduced the term "protective" to designate the foods most effective in balancing the preponderant bread-and-meat type of diet. This being the case, it is inevitably confusing to all and misleading to many if McCollum's term be now so extended as to include either or both of the very foods he sought to balance, and which still need balancing in the same directions that he pointed out.

[2] Work of Osborne and Mendel at the Connecticut Agricultural Experiment Station and Yale University.
[3] Work of Hughes et al. at the Agricultural Experiment Station of the University of California.

Hence the term protective foods when used in this book designates fruits, vegetables, and milk, the latter including cheese, cream, and ice cream.

Table 7, adapted from the work of Stiebeling and Phipard (1939) shows the extent to which several different food groups contributed to different factors of the nutritive values of the diets of 26 East North Central families, who spent from $1.88 to $2.49 weekly per capita for food. Columns for iron and thiamin are here omitted because the introduction of Enriched bread with its thiamin and iron additions has changed and is probably still changing the relative contributions of thiamin and iron from the breadstuffs *versus* the other food groups.

In the following pages, the different food groups are taken up, in essentially the same sequence as in Table 7, for discussion of their nutritional characteristics.

Breadstuffs and Other Grain Products

Bread is still the staff of life in the sense that the majority of the world's people nourish themselves more largely upon the grain crops than upon any other group of foods. Among the grains, rice predominates in the Orient, wheat as human food in the Western World. For climatic reasons, the more northern countries of Europe may depend as much upon rye as wheat; while in the United States and a few other countries, the corn (maize) crop, intended chiefly for the feeding of farm animals, may exceed the wheat crop in amount.

Regarded as food for man, wheat is "the golden grain" to the peoples of the Western World, and generally the land which suits it best is devoted to its culture, the other Occidental grain crops taking in this respect a more marginal place, even though their acreage is larger in some countries.

So widespread was man's devotion to wheat, even in prehistoric times, that the origin of wheat culture is uncertain. Clearly, however, it came into special prominence in the Nile Valley where the annual inundation left a silt favorable for

Proportion of Nutrients Furnished

Food Group	Proportion of money spent (Percent)	Calories (Percent)	Protein (Percent)	Calcium (Percent)	Phosphorus (Percent)	Vitamin C (Percent)	Riboflavin (Percent)	Vitamin A value (Percent)
Breadstuffs, cereals	18.1	31.3	28.2	11.2	20.2	1.4	5.5	4.1
Sugars, sweets	3.4	10.7	0.1	0.8	0.2	0.7	0.0	...[a]
Milk, cheese, ice cream	12.1	10.0	17.4	64.0	28.5	5.3	35.4	13.1
Butter, cream	4.9	6.0	0.3	0.9	0.5	0.0	0.2	9.3
Other fats (including fat meats)	5.6	12.7	1.3	0.4	0.8	0.0	0.7	...[a]
Lean meat, fish, poultry	24.2	11.6	31.8	2.2	21.8	0.8	28.6	7.5
Eggs	5.4	2.5	7.5	4.2	6.3	0.0	8.0	8.0
Citrus fruits	2.2	0.7	0.3	1.9	0.6	19.1	2.1	0.3
Other fruits	6.0	3.6	1.0	2.6	2.4	21.0	4.4	4.6
Potatoes, sweetpotatoes	2.2	6.1	4.6	3.7	8.3	23.7	7.7	11.2
Mature legumes, nuts	0.9	1.6	3.0	1.9	3.7	0.0	0.6	0.0
Leafy, green, and yellow vegetables	5.1	1.1	2.1	3.5	3.1	15.0	4.3	32.8
Tomatoes	1.7	0.3	0.4	0.5	0.6	7.8	0.6	8.4
Other vegetables	1.8	0.7	0.8	1.6	1.3	5.0	1.0	0.4
Miscellaneous	6.4	1.1	1.2	0.6	1.7	0.2	0.9	0.3

growth of wheat, and also helped by tending to drown out the sod grass which elsewhere handicapped wheat culture until the development of modern farm machinery.

So long as wheat was ground between stones, the different parts of the grain were more or less pulverized together except, of course, that the fibrous coat in part tended to hold together as bran particles. With the brown particles of bran removed by sifting, the resulting flour, while called white, was really cream colored and, on close examination, slightly specky because of the tiny particles of brown bran, small enough to pass through the bolting cloth.

Around 1885 mill stones were rapidly replaced by steel rollers in the grinding of wheat. The roller process is so regulated that the grain is cracked and its pulverulent inner part rubbed off as fine white flour with very little disintegration of bran or germ into particles small enough to pass into the flour. In the miller's sense this was a cleaner separation than when the grain was stone-ground. The baker also preferred the roller-process flour because it was less subject to spoilage and also made it easier for him to get uniformly the highly porous texture of the superlatively light loaf. This whiter and lighter product was, however, nutritionally impoverished. The chief reason patent flour "keeps so well" is, as R. R. Williams has cogently emphasized, because it does not afford adequate nutrition for even the lower forms of life. From the standpoint of nutritive values, the white bread of the first third of our century was in need of improvement. When thiamin became available in large quantities at low cost, a partial restoration of milling losses was effected in the form of white flour or bread, enriched by addition of thiamin, niacin, and iron.

Another nutritional improvement of white bread has been the use of increasing amounts of milk powder (for economic reasons, dried *skimmed* milk) in breadmaking. In recent years such milk powder has frequently been used in the proportion of 6 percent of the weight of the flour. The resulting improvement in the nutritive value of the bread is readily demonstrable in feeding experiments; and the opinion has been expressed that with this improvement the proportion of food calories furnished by breadstuffs and other grain products might if desired be increased to 40 percent, from the 30 to 38 percent level recently customary with unimproved bread (Sherman and Pearson, 1942).

The energy value of bread, 75 Calories an ounce, is not appreciably altered by the incorporation of skim-milk powder into the dough; by enrichment with thiamin, niacin, and iron; or by the use of flour milled in different ways. On the other hand, the nutritive value of bread protein is materially enhanced when milk powder is used, or when wheat germ is returned in the breadmaking, or when whole wheat rather than white flour is used, but is not thus enhanced by the use of enriched flour fortified by additions of iron and synthetic vitamins. The proteins of white flour when taken alone have relatively low nutritional efficiency, but even a small proportion of the protein of milk so supplements the flour proteins as to give good nutritional value to the whole protein mixture. Emphasis should also be given to the fact that the nutritive value of the protein mixture of whole wheat is much higher than that of white flour. Hence in considerations of white bread vs. brown, it is misleading merely to say that one actually assimilates about the same amount of protein from the two: one should also say that the protein of the brown bread (or of the white bread made with a liberal proportion of milk powder) is of decidedly higher nutritive efficiency than the protein of ordinary white bread.

We actually derive a larger proportion of our food protein from bread and cereals than most people probably realize. Thus the people whose dietaries are averaged in Table 7 obtained 28 percent of their protein from this food group; and in the larger number of observations spread over a longer time which have been summarized elsewhere,[4] the breadstuffs and other grain products furnished 37 percent of the total food protein.

Both in the presumably representative families of Table 7 and in the broader sampling of our population just mentioned, this food group cost only 18 percent of the food expenditure, yet it furnished in the two averages, respectively, 31 and 38 percent of the calories; 28 and 37 percent of the protein; 20 and 30 percent of the phosphorus; and doubtless at least 25 percent of the iron of the total food supply. Thus this is an outstandingly economical food group, and correspondingly it tends to occupy a larger place in the dietaries of the low-income families. It is, however, a mistake (even if a common one) to give bread only such a place in our food habits as the necessity for economy compels. Regardless of price relations, bread deserves a substantial place in the food of practically everyone. In addition to its direct nutritive values when made by the better of present-day methods, bread is also a good vehicle for several other foods (including vegetable cooking-water which too often otherwise goes to waste), and the texture of bread tends to give good physical properties to the food mass as a whole in the digestive tract.

When enriched flour and bread were being planned, two bases for fixing the extent of the enrichment (in the particular nutrients to be included) were considered somewhat fully (a) by the proponents of the plan, and (b) by the Federal Food and Drug Administration who have the chief responsibility for the legal enforcement of any such plan that may be adopted. One proposed basis was "the philosophy of restoration," the other was that the enriched food shall "carry its share" of the particular nutrients in question. The Food and Drug Administration decided that the restoration principle is "too vulnerable" because wheat does not have a quantitatively constant natural composition, and also because a law court might hold it a too

---

[4] Sherman, 1933, *Food Products*, p. 554; 1941, *Chemistry of Food and Nutrition*, p. 510.

anthropocentric view to regard Nature's means of reproducing wheat as if it were created with specific reference to the needs of human nutrition. It was argued in behalf of the restoration principle that while the wheat kernel evolved with reference to the reproduction of its species, our species has evolved in nutritional adjustment to its natural foods. But the Food and Drug Administration held that the project would fare better at law if based "realistically" upon human needs as now known and staple foods as they now are.

Scientific testimony was taken as establishing the two legally essential facts: that larger intakes of certain nutrients would be to the interest of the public health and that white flour and bread are suitable vehicles for such enrichment of the nutritional intake. Hence, it was decided that those who wish to do so may manufacture and sell enriched flour or bread, and any product thus distinguished must carry its share (as established by governmental authority) of the amounts of thiamin, niacin, and iron that are needed in human nutrition. The concept of "carrying its share" was then brought into quantitative form on the basis of relation to calories, because in general it is chiefly the energy need that determines the amount of food consumed. So as much enriched bread as covers, for instance, one fourth of a man's daily need for calories is also to cover one fourth of his daily need for thiamin, the latter being taken at the minimum-requirement level so that it may confidently be expected to be fully upheld in any court of law. The interested reader will wish to keep also in mind the fact that the enrichment program touches only some and not all of the nutritive values in which white flour has been impoverished. For some other essential factors we must depend as before upon balancing the grain products by "enough of the *right kinds*" of other foods.

Sugars and Other Sweets

Sugar in recent normal times is an even cheaper source of calories than is bread, but, even more than does bread, it needs balancing, for in respect to nutritive values, sugar is a fuel food *only*. From the viewpoint of meal planning, a few other foods are so sweet as to be bracketed with sugar and it is to their other ingredients that the fractional figures in the protein, mineral, and vitamin columns of the line for the sugars-and-sweets group in Table 7 are due.

Milk and Its Products

The different forms of milk resemble each other in nutritional character more closely than does any other food approach any of them.

Milk in its various forms is the most effective of all foods in supplementing the breadstuffs and other grain products. Recent advances of nutritional knowledge give added cogency to the time-tested principle that "The dietary should be built around bread and milk." And the same advice is an equally good guide for the planning of a national food supply or food-production program.

Greater prominence of milk in the individual dietary or the family, community, or national food supply means a more digestible diet of better-balanced protein, mineral, and vitamin content, richer in calcium, and practically always also in its riboflavin content and vitamin A value. It is also to be remembered that the three factors just named are the ones which we know with most certainty to possess large margins of beneficial increase above the merely minimal-adequate level. In other words the best amount of these nutrients is much higher than the strictly necessary amount.

The great majority of American dietaries or family food supplies, at all economic levels, can be nutritionally improved, more certainly and usually at less cost than in any other way, by increasing the proportion of milk. This need not necessarily mean taking the milk as a beverage, although to sip it little by little throughout the meal is perhaps from the standpoint of digestion the best of all ways to take a glass (or more) of milk. Used in cooking there is usually good conservation of the nutrients of milk, as there is also in the commercial preservation of milk in canned or dried forms. Other ways of taking milk are as cream or ice cream, or as any of the many varied and interesting forms of cheese.

While it is easy to see from Table 7 that milk is, both absolutely and relatively, an important contributor of calories, protein, calcium, phosphorus, riboflavin, and vitamin A value to the food supply, it is not simply upon such analytical data that we base the above statement of the importance of increased milk consumption. This latter has been directly demonstrated by actual feeding trials with schoolchildren, whose physical and mental growth has been shown to be benefited by an added allowance of milk. This has also been demonstrated in numerous well-controlled experiments with laboratory animals, often extending through entire lifetimes or even through successive generations. Noteworthy among the experiments with children were those of H. C. Corry Mann whose findings, as summarized by the careful English physician Sir Walter Fletcher, showed that even with a boarding school food supply "medically adjudged to be sufficient for healthy development, the boys were in fact not attaining to the physical and mental growth of which they had the potentiality, and to which they did attain when given an extra daily ration of milk." The British Milk-in-Schools Scheme has shown similar benefits from the addition of milk to the hitherto acceptedly normal dietaries of large population groups of schoolchildren. The added milk supported better growth in both height and weight, better general "condition" or fitness, and greater alertness of attitude and buoyancy of spirits.

Butter and Other Fats

While cream is often, as in Table 7, bracketed with butter for obvious reasons, yet nutritionally it should also be kept in mind that cream is mostly milk. It is simply a part of the milk into which has been gathered (by gravity or by a centrifugal separator) an increased share of the milk fat. Average cream with 18 to 20 percent fat is about five sixths milk, so people who find irksome "the melancholy mildness of milk" may be persuaded to take the milk nutrients in the form of cream instead. If they want to increase body weight, the extra fat in the cream makes gain of body fat so much the easier; while if one wishes to keep down one's weight, the fat of the cream may be offset by eating less of butter or of sweets.

Regarded as sources of energy in the body, fats are, weight for weight, a little more than twice as concentrated fuel as are the sugars. (On the other hand, sugars are slightly quicker fuels than fat—a fact that has probably been emphasized somewhat beyond its scientific importance.)

When a person of normal physiology subsists upon a diet of familiar staple foods, his appetite (the amount that he feels inclined to eat) tends to approximate his energy (calorie) requirement: otherwise there would be more excessively fat or excessively thin people than there are now.

Much the largest part of the day's supply of food calories is usually consumed at two or three meals. Hence one or more of the daily meals might seem burdensomely bulky unless we were able to take a considerable share of the needed calories in the concentrated form of food fats. The inclination arising from this reason is reinforced by the further fact that an acceptable form of fat helps the breadstuffs and vegetables of a meal "to go down agreeably." The facts in the two preceding sentences may be physiologically or psychologically connected.

There are also other reasons for attributing importance to fats in the food supply. They have much to do with the attainment of the flavors and textures customarily desired in some types of cooked food. Moreover, food containing a high proportion of fat, either naturally or because of having been cooked in it, tends to stay longer in the stomach and thus to make it less likely that one will feel the "hunger pangs" of an empty stomach before the next meal. It is only when sitting down to an abundant meal that hunger is a pleasure: if one is racked with anxiety as to how the next meal is to be had, the sensations arising from an empty stomach are not pleasant. These two facts—the difficulty of obtaining the desired effects in cookery without the accustomed amount of fat and the fact that a meal containing but little fat may mean hunger pangs and accentuated anxiety before the next meal—were found to constitute real hazards to morale in the First World War. Hence the prominent attention given to fats in food-supply considerations ever since.

The level at which the supply of food fat must be maintained in order to meet consumer demand is largely dependent upon individual, community, and national or racial customs. The scientific studies of the situation which had developed by the third year of the First World War indicated that for the European countries it was important to morale and efficiency to provide at least 50 pounds per year of total food fat. Yet no such dependence upon fat has appeared among the Japanese either then or since. From the beginning of the Second World War the British gave careful attention to the conservation of their food fats with the result that the lowest level of rationing found necessary has been 8 ounces per person per week for total table and cooking fats including not more than 2 ounces of butter. To this 26 pounds per year of "visible" fats (fats purchased as such) there is added the probably about equal amount of "invisible" fat contained in other foods such as meat, eggs, cheese, and milk, bringing the total fat consumption to about 50 pounds per person per year—the same as the amount deemed to be a safe minimum by the scientific advisers of the First World War period. Up to that period of the Second World War when the United States and Great Britain began to make common cause in the matter of food supply, we Americans had been consuming, in the economic sense of disappearance, about 100 pounds of fat per person per year. Some of this was probably sheer waste, as judged from the promptness of the initial retrenchment which followed the first appeal for conservation of fat in this country. To what extent there was also a voluntary shift toward partial substitution of other things for fat in the food actually eaten is not known. It is clear, however, that if actual consumption of visible fat were at the same low level here and in Britain, our situation would still be more comfortable than theirs for two reasons. We have larger per capita supplies of meats, eggs, and milk, all of which help as does butter to make bread and bulky vegetables go down pleasantly; and also in these foods we have a larger supply of invisible fat than the British food supply is likely to contain for some time to come.

Probably in all countries the invisible fat of the natural foods eaten suffices to furnish the very small amount of "nutritionally essential fat or fatty acid" that human nutrition requires. From the viewpoint of nutritional values, it is unfortunate to bracket together the fats which are, and those which are not, naturally rich in vitamin A. Yet it is difficult to make a separation in a general grouping of foods, because oleomargarine may or may not be fortified with vitamin A. When fortified, it is usually to a level near the minimum of the normal range of vitamin A value in butter, or about half as much as that of good butter from well-fed cows.

Meats, Including Poultry, Fish, and Shellfish

The average American family spends about one fourth of its food money for meats including poultry, fish, and shellfish. In return, this food group furnishes, as may be seen from Table 7, about a proportionate part of the protein, phosphorus, and riboflavin, while it contributes only one third to one half of its corresponding share of the calories and the vitamin A value, and only negligible proportions of calcium and of vitamin C.

Meats also contribute significantly to the dietary iron and thiamin (probably also some of the less well settled of the B-vitamins) but here one is probably not justified in attempting quantitative statements, partly because of uncertainties as to the accuracy of present methods when applied to attempts to compare some of the vitamin values in different types of foods, and partly because the enrichment program is resulting in larger proportions of the iron, thiamin, and niacin of the food now being contributed by the breadstuffs.

As has been emphasized by Hoagland and his coworkers in the United States Department of Agriculture, and by Elvehjem and coworkers at the University of Wisconsin, pork is richer in thiamin than are other meats; but recent studies of Hughes and others at the University of California indicate that this difference is hardly as large or as constant as seems to be generally supposed.

As meats are well liked by most people, and are traditionally associated with feasting and a high standard of living, they are commonly given a "main dish" position in the meal plan and to them is devoted a larger share of the food money than corresponds to their contributions to the nutritive value of the diet as evaluated on any scientifically valid quantitative plan that the present writer has met or been able to conceive. It is true that typical lean meats have been found to contain more of some of the vitamins than previously supposed, and also to contain some of the newly discovered vitamins. But these facts neither belong so exclusively to meat nor constitute a sufficiently large part of the total nutritional picture or accounting as to bring the contribution made by meat to the nutritive value of the diet abreast of its share of the cost. It is noticeable, too, that in the very extensive current advertising of meat, while its nutritive value is of course presented as favorably as possible, the appeal to the consumer is quite as often and as largely on traditional as on scientific grounds. Of course the scientist recognizes that, "apart from all questions of nutritional need, eating has an immense vogue as an amusement," and that various kinds of satisfaction may combine to justify an expenditure. Yet in the case of any food for which large fractions of family income are expended, one naturally looks for correlations between cost and nutritive value. Several researches upon the nutritional significances of substances known to be contained in meat and their bearings upon the place of meat in the diet are now in progress. Whether there is anything else involved which is in any sense nutritional will probably remain an open question until there have been experiments with large numbers of well-controlled laboratory animals fed different proportions of meat in otherwise identical and fully adequate dietaries throughout entire lifetimes and successive generations, such as have been made with different proportions of milk in the diet. Such experiments should be undertaken only by laboratories prepared to devote to them especially experienced direction and relatively large resources for at least four years.

Eggs

What has just been said of the need for comprehensive, long-term experimental research to determine the nutritionally best quantitative place of meat in the dietary is true also for eggs.

Eggs are relatively rich in protein of high nutritional efficiency, in phosphorus, in riboflavin, and in vitamin A. They also contribute appreciable amounts of calcium to the dietary. But they do not contribute (as milk and most fruits and vegetables do) to the body's alkaline reserve or to the maintenance of a good intestinal hygiene.

Eggs are good contributors of dietary iron and thiamin; and have been definitely shown to have high pellagra-preventive value. It may be true of eggs, as seems certainly true of milk, and probably of some vegetables, that their actual value in prevention of pellagra is considerably higher than the published estimates of niacin content would indicate; whether because the methods for niacin assay are not accurately applicable to all foods, or because pellagra as it actually occurs is less purely a niacin deficiency than current ways of speaking and writing imply, or in some less direct way, is not yet clear.

As may readily be seen from Table 7, eggs are, when compared with other foods in relation to their calories, rich sources of protein, phosphorus, riboflavin, and vitamin A value; and fair sources of calcium. Eggs also have a vitamin D value which is relatively high compared with other staple foods and undoubtedly may be of real nutritional significance, but is not high compared with that of the fish liver oils.

Fruits

The Government grouping used by the Nutrition Division of the Federal Security Agency, and also by the United States Department of Agriculture on one of whose publications Table 7 is based, distinguishes between citrus and all other fruits. Assuming that the data of Table 7 are representative, citrus fruits contribute practically as much vitamin C to our food supply as do all other fruits combined, while involving only about one third as much cost and one fifth as many calories. Citrus fruits also contribute relatively more riboflavin, while other fruits contribute more vitamin A value. Not only are citrus fruits relatively good contributors of calcium, but in addition it has been shown that orange juice specifically aids calcium assimilation.

During the period between the two World Wars, and especially in the later years of that period, there have been large plantings of citrus-fruit trees in California, Florida, and Texas. As a result, production has been increasing and, except for costs of transportation and retailing, prices of these fruits in season, and for canned grapefruit and its juice at all times of the year, have decreased to levels at which these foods with their high vitamin C content and other nutritional virtues have ceased to be luxuries and are now to be regarded as staple foods and good investments in nutritive values. This fact deserves to be more widely known and more fully appreciated. With the large areas of citrus orchards that are now coming into bearing, and the technological advances in economical preservation of these fruits and their juices, per capita consumption of them may well increase many-fold.

Beside being the outstanding sources of vitamin C, and good sources of the other water-soluble vitamins and the mineral elements generally, citrus fruits and their juices are great helps to the maintenance of a good hygiene of the digestive tract and of the body's alkaline reserve. Thus, a citrus fruit or juice is mildly acid in the mouth but after absorption into the system the citric acid is burned and the basic elements of the fruit or its juice are largely added to the body's assets of reserve alkalinity. This continues to be true even when one consumes up to four pounds of the fruit or four pints (eight full glasses) of the juice of oranges or grapefruit a day (Blatherwick and Long, 1922; Lanford, 1942).

The tomato, which is a fruit although often treated as a vegetable, is in normal times widely available (fresh or canned) and in some regions may be significantly cheaper than orange or grapefruit—though this is now by no means so often true as formerly. Tomatoes are less rich in vitamin C than grapefruit, and only about half as rich as oranges, but are nevertheless an excellent source of vitamin C; and they are richer in riboflavin and vitamin A value than are oranges or grapefruit. Both tomatoes and citrus fruits hold their vitamin C value excellently in canning and storage.

Other fruits are not quite so fortunate in this respect but still tend to retain their vitamin C values about as well as the general property of freshness is preserved. It is also a fortunate fact that fruits as well as succulent vegetables usually have their maximum vitamin values at about the stages of maturity at which we most enjoy eating them, and at the times of year when each is at its best and at the height of its season and therefore most abundant and least expensive.

As different cultural varieties of the same species of fruit have been developed for the sake of differences of flavor, and also of such diverse properties as color, earliness of maturity, keeping and marketing qualities, it is easily conceivable that as an accidental and unforeseeable accompaniment of the development of these other qualities, considerable differences in vitamin value may have been introduced which are only now coming to light. The best-known case of this kind is that of the apple, hundreds of varieties of which have been developed; among them perhaps a score or more have now been studied as sources of vitamin C with the result that some varieties are found to be on the average perhaps 3 times as rich in this vitamin as some others. A case of the opposite kind is that of the banana, of which the trade decided to promote only one variety in the American market. So taking all varieties of apples in our market as they come, we find apples *averaging* about the same but *varying more* than bananas in vitamin C value. Thus, if one selects for himself a single variety of apple, then according to the variety chosen it may be either richer or not so rich as the single variety of banana which the growers and distributors selected for the American public.

The mild acidity of most fruits and the texture of the flesh fiber of many of them combine to make raw fruit an especially pleasant and wholesome food with which to end a meal, because these properties leave the teeth and gums in a clean, fresh condition, and also aid the digestion of the meal as a whole. Hence, as often as one can conveniently arrange to do so, let the last course of each meal be not anything sticky, pasty, or artificially sweetened, but rather a fruit salad, an apple-celery salad, or a raw apple. And if any food is taken between meals, give strong preference to raw fruit or fruit juice.

Vegetables

It is. of course, as staple vegetables of high fuel value that potatoes and sweetpotatoes are bracketed as a separate food group in Table 7. These two foods are not closely related botanically: one is a tuber, the other is a root, and they belong to plants whose scientific classification is not close—for this reason science seeks to "take the curse off" an illogical name by writing sweetpotato as a single (and thus obviously artificial) word. They are, however, not very different in most features of their nutritive values; and the food habits of the American people treat them as interchangeable electives in meal planning. The starchiness of both, plus the sugar of the sweet-potato, makes them relatively high energy foods. They do not contain high percentages of protein (even if figured on the basis of dry solids), but potato protein has been shown to have high efficiency in human nutrition, while sweetpotato protein remains to be investigated in this respect. On the other hand, potato has a relatively low, and sweetpotato a relatively high, vitamin A value. Both become fairly important sources of thiamin and riboflavin when used in the quantities that American food habits make practicable. Both are also fairly good sources of vitamin C when cooked in their skins and eaten without undue delay.

Mature legumes and nuts constitute another food group in the United States Department of Agriculture classification followed in our present Table 7. While true nut devotees might consider such a bracket incongruous, the peanut bridges the gulf between the "lowly legumes" and the "true" or tree nuts; and peanuts constitute an increasingly important food crop, while other nuts at present are only a minor (though interesting) feature of our food supply. The legumes including peanuts are protein-rich foods, and peanut protein has been shown to have high nutritive efficiency and to be so constituted as excellently to supplement the protein of bread. Studies of the nutritive efficiencies of the proteins of other legumes, including our ordinary beans and peas, have yielded results so variable as to be difficult of interpretation. Some feel that because of occasional indications of a lesser efficiency in the legume proteins, the traditional discrimination in favor of "animal proteins" is confirmed; while others, including the present writer, consider the evidence to be so generally favorable to legume-protein as to justify the view that legumes and nuts may be regarded as nutritionally satisfactory alternatives to meat for those who may wish so to use them either occasionally or often, and whether for reasons of palate or purse.

Tomatoes, while separately placed in Table 7, are usually

bracketed with citrus fruits for nutritional consideration, and are so discussed above.

Green and yellow vegetables as a group are important for their contribution to the vitamin A value of the diet; but science *does not* specifically seek the sanctification of spinach! In fact spinach is now known to be an unfortunate choice among the green-leaf vegetables because it contains a relatively large amount of oxalic acid, which is not a desirable substance for human consumption in any case and which renders practically unavailable and useless the calcium which spinach, chard, and other leaves of the Goosefoot family contain. On the other hand, many other green foods, including broccoli, cabbage greens, collards, dandelion, kale, loose-leaf lettuce, turnip tops, and watercress are practically free from oxalic acid and are important dietary sources of calcium as well as of vitamin A and riboflavin.

Carrots, like sweetpotatoes, are rich sources of vitamin A values, though their calcium contents are much like those of other roots. The carotenes which give these vegetables their yellow color and vitamin A value are also contained in important amounts in yellow corn; but not in yellow turnips, whose coloring matter is of a different kind.

Broccoli is a vegetable of high nutritive value and deserving of its increasing popularity. Formerly regarded as a somewhat exotic luxury, it is now quite cheap, especially as one may enjoyably and with nutritional advantage include in the edible portion not only the flower bud but also the adjacent leaves and the tender green twigs on which these are borne. These twigs are comparable with the edible stem and tip of asparagus. The leaves are among the richest of foods in calcium, iron, and vitamin A values and are also good sources of other vitamins. The flower buds constitute, of course, the distinguishing feature of broccoli as a food, and far outrank the one other flower we commonly eat (cauliflower) in colorful character and nutritive values. Broccoli is a relatively sure crop, easy either to market or to raise in home gardens. To the consumer who grows his own it offers the great practical advantage that the same plant continues to grow successive pluckings of the edible portion throughout a very long season.

A Larger Grouping

Milk and its products, fruits, and vegetables collectively constitute the part of the dietary or food supply to which our present-day knowledge gives increased attention and prominence. How much of our food expenditure do we invest in this collective group, and what proportion of our nutrient intake does this investment bring us? Or alternatively, what proportion of our total calories do we draw from this larger group of foods, and what percentages of our other nutrient intakes do these foods bring us along with those calories?

These are, of course, two ways of putting the general question of the nutritional characteristics of this larger food group as compared with the other foods collectively or the dietary as a whole.

Here again, the data of Table 7 may be taken as fairly representative. In this larger group (milk and its products, fruits, and vegetables) the families represented in Table 7 invested (in nearest whole numbers) 37 percent of the money that they spent for food, and for this fraction of their food money they received from this group of foods 30 percent of the calories, 30 percent of the protein, 81 percent of the calcium, 49 percent of the phosphorus, 97 percent of the vitamin C, 56 percent of the riboflavin, and 80 percent of the vitamin A value. Thus, viewed in the light of the investment in them as compared with the dietary as a whole (or with all other foods of the dietary collectively), it is clear that this group (fruits, vegetables, and milk) contributes outstandingly to the mineral and vitamin values, and at the same time brings to the dietary almost its *pro rata* share of the calories and protein.

Or, characterizing on another basis, we may say that to the extent that the families represented in Table 7 drew upon this food group toward the meeting of the calorie requirement these foods furnished with their calorie quota, their exact protein quota also, and at the same time much more than their quotas of the minerals and vitamins here considered. And if protein instead of calories were made the basis of this latter comparison, the same thing would appear as to the high mineral and vitamin values of this collective group of foods.

Thus from whichever of these viewpoints we characterize the food group consisting of milk and its products, fruits, and vegetables, these foods collectively stand out clearly as the effective way of bringing into the individual dietary or the national food supply that more generous allowance of mineral elements and vitamin values which will certainly improve our current American practice.

CHAPTER X

"ARE WE WELL FED?"

THE title of this chapter is borrowed from the essay by Dr. H. K. Stiebeling published by the United States Department of Agriculture in 1941 as its Miscellaneous Publication No. 430. The opening words of that publication are:

Strong and alert nations are built by strong and alert people.

Strong and alert people are built by abundant and well-balanced diets.

No nation achieves total strength unless all of its citizens are well fed. To be well fed means more than filling the stomach with foods that appease hunger. It is more than getting the food that barely protects the body from disease due directly to poor diet. It is having each day the kind of food that will promote abounding health and vitality.

Our Nation's goal is that everyone shall have a diet adequate in every respect for good nutrition.

As compared with most countries, the United States has rich and varied resources for food production. Yet the food-consumption studies of 1936–37 showed a large proportion of our families to be subsisting on diets which nutritionally could not be judged good; and the draft examinations of 1940–41 showed a high percentage of our young men to be disqualified by physical impairments attributable in greater or less degree to the fact that so many of these men had not been well fed.

In Britain it is an open secret that when a presumably fair sample of similarly rejected men were given good feeding for a few months a very large majority of them were rendered fit.

From the United States Public Health Service comes the opinion that nutritional deficiency diseases "in all probability constitute our greatest medical problem, not from the point of view of deaths, but from the point of view of disability and economic loss." A little reflection will make it abundantly plain that the "economic loss" of the preceding sentence stands also for many frustrations and losses of satisfaction in life.

Undoubtedly too, malnutrition plays a much larger part in our death rates than our vital statistics reveal. A committee, two thirds medical, appointed by the National Research Council to study the prevalence of malnutrition in the United States, found no way of obtaining numerically accurate data from the official records of deaths, but reported the existence of indubitable evidence that malnutrition is a much larger factor in both morbidity and mortality than has hitherto been supposed (Jolliffe, McLester, and Sherman, 1942).

This committee considered it probable that deaths actually due to malnutrition in the United States are many times greater than the mortality statistics indicate; and that data based on hospital records are subject to the same criticism as the official mortality rates, namely, that they are incomplete because of (1) nonrecognition, and (2) mislabeling in the traditional "precedence" given to certain diseases over others. Thus it is still impossible to know how much malnutrition is concealed, for instance, in the 370,000 deaths recorded in 1938 under the heading of diseases of the circulatory system, or how much under "senility," "cirrhosis of the liver," and "psychoses."

Even among physicians thoroughly attuned to the importance of the nutritional status of their patients, recognition of early stages or subacute cases must depend upon the delicacy of the diagnostic methods available at the time. With a method more delicate than any previously in use, Kruse (1941) and also Jolliffe (1942) find evidence of a much greater prevalence of mild shortage of vitamin A in New York City than had previously been imagined.

Very important also is the evidence of widespread nutritional handicap which came to light when at the Toronto General Hospital one half of a large group of expectant

mothers were given the usual advice and supervision of the
antepartum clinic, while the others were given in addition
a daily supplement to their accustomed home diets. This daily
supplement consisted of one egg, 30 ounces of milk, 0.5 ounce
of wheat germ, 1 ounce of cheese, 4.5 ounces of canned tomatoes, and 1 orange. The mothers who received this extra food,
and also the children to whom they gave birth during this
experimental feeding, made strikingly better records than the
parallel patients who continued on their accustomed home
diets.

Dr. Jolliffe's committee concluded that strikingly obvious
malnutrition is relatively rare in the United States, that cases
apparent only to the physician who looks for them are frequent, that the cases which are vague and elusive even to the
skilled and careful observer are probably more frequent still,
and that "if optimal nutrition is sought, not mere adequacy,"
then there is need for widespread improvement in the food
of the people of the United States.

The incompleteness of reporting of malnutrition thus shown
makes it probable that the findings of food-consumption
studies are more significant from our present viewpoint than
any vital statistics or clinical records yet available for most
or all of the United States.

As Stiebeling points out, the food-consumption data indicate that millions of people in this country are living on
diets that are below the safety line. This does not mean that
all these people are consciously hungry (though "some often
are"). "Nor do all that subsist on poor diets show symptoms
of pellagra, beriberi, scurvy, anemia, or other well-defined
disease."

Beside conscious hunger there is the so-called "hidden
hunger" of the people who have enough of total food but not
enough of the "right kinds of food" and these latter are, in
the United States, usually much more numerous than the
former.

Our greatest nutritional handicap in the United States is

not that part of our population which is starving in the historic sense, nor that part which is *recognized* as suffering from specific nutritional deficiency diseases, but the part (probably much larger than those other two parts put together) which is "getting along on poor diets." Most of the individual handicaps of this latter kind or degree go entirely unrecognized or at most are regarded as "merely borderline cases"—a phrase which seems to cast doubt upon the reality of the affliction, whereas what is more probably doubtful is the adequacy of the method of diagnosis.

With more delicate methods of diagnosis many of the "borderline cases" are recognized as "frank" cases of one or another nutritional deficiency disease. But how much better still if they were entirely prevented by good feeding.

Under the rude awakening of the draft examinations and the stress of war, it was recognized that, as Dr. Stiebeling put it: "The Nation's families need good diets to safeguard their own health and to strengthen the defenses of the country."

Our permanent policy should be not merely "to find where the disease exists and cure it," but rather to safeguard against any occurrence of dietary deficiency disease in this country where our resources for food production are so ample.

At present, as the investigations of Stiebeling and coworkers clearly show, the nutritional quality and safety of diets in the United States varies greatly. In the above-quoted summary of these investigations, diets which contained less than the stated minimum [1] of any of the following nutritive values—protein, 50 grams; calcium, 0.45 gram; phosphorus, 0.88 gram; iron, 10 milligrams; vitamin A, 3,000 International units; thiamin, 1.0 milligram; vitamin C, 30 milligrams; riboflavin, 0.9 milligram—were rated "poor" or as failing to safeguard. This was considered as analogous to an "unsafe" rating on a bridge: not meaning certain disaster, but absence of a due margin of safety in one or more respects. Diets which met this 8-point

[1] These "minimum" values are lower than the Recommended Allowances of the National Research Council's Committee.

minimal standard in all respects, but with less than a 50 percent margin, were rated "fair." Those that were 50 percent
(or more) above these same minima at all 8 points were rated
as "good." As a further indication of the significance of "good"
as contrasted with "fair," this official publication states:
"Laboratory experiments and human experience indicate that
proper diets not only can lengthen the entire span of life but
that they also can lengthen the active, fruitful period, post-
poning the effects of advancing age. They can make old age
itself more healthful, less a period to be looked forward to
with dread, and less of a burden on society."

Stiebeling points out that four facts should be kept in mind
in connection with interpretation of the data of the food-
consumption studies: (1) The tentative nature of present
knowledge of the mineral and vitamin values of many foods,
especially with reference to losses in cooking. (2) Lack of
knowledge of household wastes of edible food. (3) The fact
that each diet study covered only a single week. (4) Lack of
completeness and precision in our knowledge of our nutritional needs. "Were the same diet records to be evaluated at
some future date when such knowledge is more advanced,
it is probable that somewhat different conclusions would be
reached. Nevertheless, the comparisons made among different
groups of families are significant because the diets of each
group were analyzed in the same way."

Of the families whose food-consumption data were collected
in 1935–36, presumably representative of the people of the
United States, about one fourth had "good" diets in the sense
explained above; more than one third had "fair"; and another
third or more had "poor" diets.

Stiebeling finds that, as a rule, relatively more of the ill
fed are found in the lower income levels than in the higher,
more in the larger families than in the smaller, more in the
Southeast than in the North and West, more among Negro
than among white families, more in the cities than on the
farms.

The main reason why farm family diets generally are better than those of village or city families is that the farm families eat larger amounts of fruits, vegetables, milk in its various forms, and eggs.

Of the high proportion of low-income city and village families, Stiebeling writes: "Careful planning and wise marketing are necessary if frugal food allowances are to supply all dietary needs." In 1935–36 a considerable majority of city and village nonrelief families had incomes below $1,500 per family per year. In 1942, perhaps due to "war boom," this was said to have risen to a median family income of about $2,300 a year.

"The proportion of families with diets that are good would, of course, shift with changes in income distribution and in food-purchasing power. Even without changes in the economic situation the proportion of good diets could be greatly increased if all families used their resources for food to the best advantage."

One important food resource is, of course, the home garden. Obviously farm families have a great advantage in facilities for producing a major part of the food they consume. But even a villager may raise his family's diet from fair to good by what he can easily grow in his home garden, and in his spare time.

Stiebeling puts the worthwhileness of home gardens on the right basis in the question: "Does it pay in better diets to produce food at home?"

Whether it pays in better diets and therefore in better health for all members of the family is more important than how a money accounting on the home garden project would look. Probably in terms of money the hours spent on the garden will seem to have brought but small return; but the home-garden project has other aspects which are more important than the small amount of money involved. The home garden brings all the family increased amounts of protective foods at their best and with no possibility of losses of vitamin values through marketing delays; this improvement of the diet benefits the health of all and gives the children particularly a better chance of successful and satisfying lives; and knowing these benefits to come from his culture of his garden, the spare-time home gardener will find his garden-making an enjoyably recreative form of exercise.

In many connections, and not least with respect to the way in which the home garden pays, we do well to remember what Galsworthy preached in his serious postwar mood: "Make health rather than wealth the ideal; and make that ideal effective by education from infancy up."

Under the stimulus of war conditions, a keen appreciation of home gardening has grown up in Great Britain. On a standard-sized garden allotment of 30 x 90 feet, a reasonable share of the spare time of an average family, with fair land and fair luck, produces the greater part of both the vitamin A value and the vitamin C that it needs per year, along with much else that improves the nutritive quality of the family dietary.

It is believed that, because of awakened nutrition consciousness, there is less malnutrition in Britain now than there was before the war, improved management of home resources having more than offset all the enemy depredations upon food imports as well as the war-emergency shifting of ship space to directly military needs.

In the July, 1942, issue of the Milbank Memorial Fund *Quarterly*, Kruse has contributed an interesting paper entitled "A Concept of the Deficiency States." In outline, his concept is that a dietary deficiency disease develops in the sequence: lowered concentration of the essential factor in the blood; depleted storage in the body's reservoirs; diminished excretion in the urine; microscopic change in tissue; gross morphological and functional change. One step is not necessarily completed before the next begins. Kruse has found specific biomicroscopic changes for the four diseases due to deficiency of vitamins A and C, riboflavin and niacin, respectively; and he writes that his "observations, combined with the results from administration of specific therapy, have clearly shown that the early or mild state has a deeper meaning than was previously recognized." Kruse emphasizes the distinction between acute and chronic forms of the same deficiency disease, each of which forms may be either mild or severe. Of the chronic state Kruse holds that its essence is time. The longer people live, the more chance they have to incur changes and to have them develop to an advanced state. Consequently, chronic changes are seen with greater frequency with increasing age. Kruse also holds that these chronic alterations have in the past been called senile changes, but that "senility *per se* is not responsible for them." Usually they are due to dietary deficiencies running over a period of years. Their incidence and degrees of severity vary largely with economic conditions. "Most important of all, they are reversible, yielding slowly but completely to appropriate therapy." (Kruse, 1942, p. 255).

Chronic changes, however, are found by Kruse to respond much less promptly to treatment than do the acute cases which often yield dramatically rapid cures. Kruse emphasizes the probability that chronic cases may go unrecognized and that if treated the treatment may not be sufficiently prolonged. When the possibilities of the different forms of the same deficiency condition are recognized the apparent discrepancies between frequencies of dietary inadequacies and occurrences of deficiency states are in great measure cleared up.

In the publication whose title has been borrowed for this chapter, Stiebeling concludes that, "As compared with many countries, ours is rich in food. But we still are far from being a well fed people."

CHAPTER XI

THE NUTRITIONAL IMPROVEMENT OF LIFE

NOTWITHSTANDING the high percentage of rejections for physical reasons in the national draft, the average figures show that our young men of today are about an inch taller and correspondingly better developed than were their predecessors of 25 years ago at the same ages.

Records of both men's and women's colleges have also shown that both boys and girls enter college younger yet taller than their parents and other predecessors did from 25 to 30 years ago. Though some fluctuations appear when comparisons are made over short periods, there is no reason to doubt the general trend to better physical and prompter mental development over the period of the past two, three, or four decades. The news most recently received from Yale is that the new freshman class is the youngest and the tallest yet recorded there (1943).

In none of these comparisons can the improvement be explained on genetic grounds. While increased interest in outdoor life and the teaching of other health habits have doubtless played some part, it is probable that the newer knowledge of nutrition, the growth of intelligent nutrition consciousness, and the resultant better feeding of children and more scientifically guided food habits continuing into youth and adulthood have been the chief factors in this improvement in physical and mental growth and development.

Probably, too, it was very largely because of the growing attention given to nutrition even during the first three decades of the century that the public health bore up as well as it did during the long economic depression which followed the crisis of 1929–30.

Nation-wide appreciation of the importance of nutrition grew steadily with the new knowledge in this field of science and with the increasing social awareness of the need for its further application. This was accentuated by the experiences of the economic depression and the strong feeling of unrest that there should be people inadequately fed at the same time that so-called surpluses of food crops existed. The Australian delegate to the League of Nations called for a worldwide policy which would "marry agriculture and health."

McLester in his presidential address to the American Medical Association in 1935 expressed the opinion that as science conferred on those peoples who availed themselves of the newer knowledge of infectious diseases, better health and a greater average length of life, so also it promises to those who will take advantage of the newer knowledge of nutrition, greater vigor, increased longevity, and a higher level of cultural attainment.

In 1936 Sir John Boyd Orr, doctor of medicine and director of the Imperial Bureau and Rowett Research Institute of Nutrition, wrote in his *Food, Health and Income:* "This new knowledge of nutrition, which shows that there can be an enormous improvement in the health and physique of the nation, coming at the same time with the greatly increased powers of producing food, has created an entirely new situation."

Dr. Frank Boudreau also voices the growing body of medical opinion which regards the present era of nutritional knowledge and power as something essentially new, when he says that "the benefits of an abundantly adequate diet are greater than we had any reason to expect."

The same serious attitude of realization that our generation has here encountered an unexpectedly great responsibility and opportunity for the advancement of human welfare pervades the publication of the International Labor Office on *Workers' Nutrition and Social Policy,* and the 1937 publication of the League of Nations entitled *Nutrition: Final Report of the Mixed Committee of the League of Nations on the Relation of Nutrition to Health, Agriculture, and Economic Policy.* In the former, the International Labor Office recorded its findings that large numbers of the working population, even in what were considered normal times and in the most advanced industrial countries, are inadequately nourished; and it considered what was being done to bring about better nutrition as an important factor in the public welfare. And the 1937 publication of the League of Nations definitely calls upon governments to extend their sphere of action for the betterment of nutritional conditions in the light of the new concept which had come "through research conducted over a long series of years by individuals, in a number of countries." It is significant to find a "Mixed Committee" representing health, agriculture, and economic policy basing its recommendations regarding governmental food policies so clearly upon the guidance of research in nutrition as an experimental science. In its 1937 report, this Mixed Committee emphasizes general agreement on the need for spreading a knowledge of the newer aspects of nutrition, the convincingness of the evidence of its importance to health, and the "equally abundant and striking evidence" that optimal nutritional condition is rare.

In line with the American evidence cited at the beginning of this chapter, the League of Nations' committee held that food habits in the Western World were (1937) gradually "tending to change in the right direction," the people now consuming "more milk and dairy products, more fruit and more vegetables than a generation ago." The committee considered that this improvement had resulted in part from a better understanding of dietary needs, in part from a general increase in income and in ease of transport, and in part from improvements in the methods of production and distribution of these rather perishable agricultural products. This Mixed Committee inclines to give about equal credit to the influence of agriculture in providing these perishable foods more cheaply and in higher quality; and, on the other hand, the influence of enhanced consumer demand upon agriculture. The committee considered that in 1937 the movement towards better nutrition had made considerable progress, but had not gone nearly far enough. "Poverty and ignorance remain formidable obstacles to progress." And in this report it is predicted that if adequate levels of food consumption can be achieved, it would make possible now a further progress, fully equal to that of the nineteenth century, in raising the quality of human life. For the improved health "will certainly," this committee believes, be accompanied by increased efficiency and a rise in material welfare and happiness. Hence if the hope which nutrition holds out can be made a reality, new perspectives will be opened up for the improvement of human welfare.

The League of Nations committee also held that in spite of the application of the knowledge of medicine and hygiene, the physical condition of a large proportion of people is far below what it should be, and that the more recent advances in medical science have shown that this inferiority is largely due to imperfect nutrition.

The passage above quoted from McLester was prominently reproduced in the 1939 Yearbook of the United States Department of Agriculture, which under the title *Food and Life,* was essentially devoted to nutrition. And in the Foreword of that Yearbook, Secretary Wallace wrote that probably 99 percent of the children of the United States have heredity good enough to enable them to become productive workers and excellent citizens provided they are given the right kind of food, proper training, and ordinary opportunities; but that half of them "do not get enough in the way of dairy products, fruits, and vegetables to enable them to enjoy full vigor." Wallace added that it is the duty of producers, distributors, consumers, and the Government to coöperate "to see that the children of these people are better fed than their parents were."

Nutrition was given governmental recognition in the organization of Federal defense and security activities in 1940; and at the same time the National Research Council reëstablished a standing Committee on Food and Nutrition, which has since evolved into a Food and Nutrition Board with many contributing Committees. In 1941, the President of the United States called the first National Nutrition Conference; and Vice President Wallace in his address to that Conference proposed a series of three ascending goals for the nutritional improvement of our national life:

(1) The complete eradication of nutritional deficiency diseases—"we do not have yellow fever any more in this country: we should not have pellagra."

(2) A great reduction of those infectious diseases like tuberculosis, the incidence of which depends largely upon nutritional status. And, finally,

(3) To bring within the reach of all our people such supplies of the right kinds of foods as to make it possible for all who will to attain the efficiency and enjoyment of "health plus."

The fact will bear emphasis that in the nutritional improvement of life, the life cycle becomes longer because stamina has been stronger throughout the course of the life. If the "course" of the life "cycle" be compared with the path of a projectile, the person who, for either genetic or nutritional reasons or both, is possessed of superior well-being lives on a higher plane as well as for a longer time; somewhat as a superior cannon can, if rightly aimed, throw its projectile both higher and farther.

But the actual difference that nutrition can make between, as the *Journal* of the American Medical Association editorially puts it, "passable health and buoyant health" is something much more vital than any mechanical analogy can adquately suggest. At the National Nutrition Conference above mentioned, Dr. Thomas Parran, Surgeon General of the Public Health Service, put some of the human implications as follows:

. . . given the national will to do it, we have the power to build here in America a nation of people more fit, more vigorous, more competent; a nation with better morale, a more united purpose, more toughness of body, and greater strength of mind than the world has ever seen. This total result can be accomplished only by putting to work all of the scientific knowledge we have for the nutrition of all of our people . . . , great assets in food production, distribution, education, social awareness, and patriotism can be canalized, through science, toward our goal of nutrition to lift our level of achievement.

And Dr. Boudreau has recently said that if all we now know about nutrition were applied to modern society, the result would be an enormous improvement in public health, "at least equal to that which resulted when the germ theory of infectious diseases was made the basis of public health and medical work."

Of late it has been learned (as was noted briefly in Chapter I) that the chemical specificity of a species is not so rigid but that it is possible nutritionally to induce important modifications of, for instance, the mineral and vitamin contents of the working tissues of living organisms. In January, 1941, under the directorship of Professor L. A. Maynard of Cornell, the United States Nutrition Laboratory began full operation in its investigation of the nutritional interrelationships of soil, plant, farm animal, and man. Rightly enough, there are high anticipations of improvement in nutritive values of foods through such investigations.

Even more important is the prospect of the benefit to man's own tissue composition and condition of internal environment, as people come to realize more fully and keenly the deep-seated and far-reaching effects of differences in what enters the body as nutriment. A difference in tissue composition so small as to be barely detectable (or perhaps too small to be detected) by chemical analysis may yet make a very great difference in the course of the life cycle. For, as we have seen, in the case of superior nutrition, the "course" runs both higher and farther—the more scientifically guided choice of food enhances both the quality and the duration of our lives.

Already, then, our new knowledge of nutrition has become a very important resource for each individual's constructive management of his own life history; and on the community, state, or national scale it is a powerful and far-reaching instrument of social policy, and of our agricultural, industrial, and military efficiency as a people.

The nutritional science of today with its specific facts for the prevention and cure of so many diseases, and with its further general principle of the nutritional improvability of the normal, is an outstanding new illustration of the old saying that knowledge is power.

How then is this new power to be brought into most effective use in support of our war effort, and of the constructive work of the postwar period? One form of answer to this very comprehensive question is a threefold grouping of ways to make nutritional knowledge more effective.

(1) Immediate improvement of the nutritional environment in which we and our fellow citizens live, by wholesale restoration in foods of universal consumption of nutritive values lost in processing, (that is, "the enrichment program");

(2) Education in the essentials of nutrition, the nutritive values of foods, and the influence of the kinds and relative amounts of food eaten upon the level of health and efficiency, so that people will *want* the benefits which the newer knowledge of nutrition offers; and

(3) Economic measures to bring (as Lord Astor put it) "not only enough food but enough of the right kinds of food" within the reach of all the people as consumers.

Let us consider each of these three programs in turn.

The enrichment program has so far embraced four projects dealing respectively with table salt, bread, butter substitutes, and milk.

Several years ago it began to be realized that in regions where drinking waters are nearly devoid of iodine (iodides), the people may develop an iodine-deficiency disease characterized by goiter, if they use table salt so refined as to have been robbed of the small proportion of iodide which rock salt, sea salt, and the natural brines of salt wells contain. Hence arose the custom of restoring a little iodine (in the form of sodium or potassium iodide) to table salt. The product is called iodized salt; but it need not be regarded as anything medicated, for the iodide is simply a restoration of what natural salt contains. Such restoration to table salt of a natural constituent of which it had been impoverished in refining has been largely promoted by public-health authorities and found very effective in greatly lowering the incidence of goiter.

Recently, as described in Chapter IX, the same principle is being applied in the restoration to white flour or bread of one or more of the nutritionally important substances which the modern milling process rejects. This is now being done in one way or another in several countries including our own, and in all cases the project starts with the proposal to restore thiamin (vitamin B_1). In Great Britain calcium is being added along with thiamin; while in the United States this enriched flour and bread must be restored to something approaching whole-wheat levels in its thiamin, its iron, and its niacin (nicotinic acid) content, and later is to be enriched (according to an announcement of the Federal Food and Drug Administration) in riboflavin also. In both countries the discussion of the enrichment of artificially impoverished flour and bread has centered upon its thiamin content. The fact is clear that white flour had been impoverished (in the modern milling process) of much the greater part of the thiamin which wheat naturally contains, and that this nutritional impoverishment of a food which practically all consume—most people as one of the major foods of the everyday dietary—had had the effect of bringing down the thiamin content of the diet to a level too low to support the best of health.

This fact gives great force to the argument that there should be a restoration of what the modern milling removed, or an enrichment of its impoverished product. As has been explained in Chapter IX, the Federal Food and Drug Administration decided to use the term enrichment for the project and Enriched as the designation to distinguish and identify the product. This term Enriched does not ensure that the thiamin content has been brought fully up to whole-wheat levels, nor does the use of enriched bread in one's regular daily diet necessarily mean that the thiamin problem is thereby completely solved. But if one follows the simple rule "that all white bread be enriched," the effect will certainly be to bring the thiamin contents of American dietaries (and their iron and niacin contents also) to distinctly more satisfactory levels. And we may reasonably hope that in a large proportion of cases this enrichment may, with the natural foods which the dietary also contains, bring the thiamin, iron, and niacin intakes up to approximately their optimal levels. So we believe that the enrichment program is deserving of universal support because of the benefit to our national nutritional environment which can thus be brought about in a relatively short time.

The other two enrichment projects thus far established are, respectively, the enrichment of table oleomargarines in vitamin A value, and the enrichment of the vitamin D content of milk.

By an extension to butter substitutes of the same general principle which was followed with artificially impoverished flour, it is sought to ensure the presence in table oleomargarines of vitamin A value reasonably approaching what may be considered "their share," perhaps half as much as in good average butter. Enrichment of the vitamin D content of milk —to make a good natural food better balanced—is a different extension of the same general principle.

In the judgment of nearly all nutritionists, public health experts, and food control officials, all four of the current enrichment projects are welcome improvements of our nutritional environment. They leave plenty to be done by nutrition education and by economic action to bring larger supplies of the nutritionally most desirable foods within the purchasing power of the low-income families of the United States.

Education of the public in the essentials of nutrition and the nutritive values of foods is and will be important in any case. For even if all foods which have been impoverished in processing, and all substitute foods, could be enriched up to the levels of nutritive values of their prototypes, there would still be great additional benefits to be gained through education in the use of nutritional knowledge in our everyday decisions as to what kinds of food we shall eat and in what relative amounts to eat them. The benefits to be gained through learning and practicing a more scientific use of natural foods, even in the everyday respects just mentioned, will very probably require more time for accomplishment than the enrichment program, but they extend far beyond what the enrichment program has to offer. The highest ideal of enrichment of flour would be to give it all the nutritive value of whole wheat; but whole wheat alone does not suffice our nutritional needs. One of the problems of around twenty years ago was to find what proportions of other foods, in McCollum's phrase, "protective foods," would so supplement wheat as to make an adequate diet.

Here, for the sake of conciseness, we must (so to speak) telescope the findings of a great deal of recent research. The evidence of direct human experience is combined with that obtained by laboratory experimentation extending often throughout the entire lifetimes and even successive generations of animals of species whose natural life cycle is short enough to make possible this comprehensive plan of research and whose nutritional chemistry is so related to our own that we know how to be strictly scientific in applying the data to our human problems.

Until recently it was accepted as a truism in physiology, and sometimes expounded as a principle in economics, that one cannot advantageously consume much more food than the amount that one actually needs to prevent obvious loss of body substance. This idea has now been shown to be an undue simplification, correct only for some and not all aspects of food and nutrition. Of total food as measured in calories, it is true that once one has established oneself at a normal body weight for age, sex, build, and activity nothing is to be gained by eating much more than just so many calories as are needed to maintain this status. But, as has already been mentioned briefly and will be discussed more fully later, we now have conclusive experimental evidence that nutritional well-being is enhanced by higher levels of intake of food calcium, for instance, than one strictly and demonstrably needs for maintenance of a normal calcium status. Here there is, as McCollum puts it, a wide difference between the minimal-adequate and the optimal in nutrition. This fact is established with particular conclusiveness and knowledge of detail with respect to calcium, where different levels of nutritional intake (calcium content of food) have been studied through successive generations of large numbers of well-controlled laboratory animals. Evidence of the same nature shows the same thing to be true of the riboflavin content and the vitamin A value of the diet, though these cases are still in course of being worked out in the same degree of detail as was the case with calcium. Evidence of less comprehensive kinds but more largely obtained directly from human subjects seems to indicate that similarly liberal margins of thiamin and of vitamin C intake are beneficial. But these concrete facts should not be generalized away into a loosely open-handed attitude or "principle of nutritional liberality" with respect to all essential nutrients. If one desires to speak or think in terms of "a principle" in this connection, it should be a *principle of scientific discrimination.* What level of marginal intake above that of strict necessity or minimal adequacy is most advantageous is a question which can properly be answered only on the basis of comprehensive research upon each nutritional factor independently.

It has sometimes been assumed that the optimal level of intake of any nutritional essential is that which will keep the body saturated in the sense explained in our discussion of vitamin C. While it probably is desirable to keep the body thus amply supplied with this vitamin and some others, it is an illogical extrapolation to assume that saturation is desirable in the case of all the vitamins we need in our nutrition.

The permanently desirable level can be conclusively ascertained only by lifetime experiments with different nutritional levels; or, better, experiments extending through the lives of at least two generations. By such experiments we have found that with calcium, riboflavin, and vitamin A the body continues permanently to profit by increased intakes up to more than twice the level which is directly demonstrable as necessary.

The combined evidence now shows clearly that starting with ordinary or typical American dietaries of all economic levels, the chief direction for nutritional improvement is to increase the proportion of fruits, vegetables, and milk in its various forms; and that such improvement even of dietaries already adequate and supporting a level of health already normal results in better growth and development, higher attainment in stamina and working efficiency, and a longer lease of healthier and more useful and satisfying life.

Of this comprehensive finding, the part which has probably attracted most attention because least expected is that our present knowledge of nutrition offers a longer life to the average normal man or woman who chooses to use it in his or her everyday food habits.

Obviously, however, the fullest benefit is to be had not by those who learn of this only in adult years, but by those whose food habits are so guided throughout life.

CHAPTER XII

NUTRITION FOR REALIZATION OF THE POTENTIALITIES OF YOUTH AND OF MATURITY

AWARENESS of the importance of right feeding of infants and growing children was both an early manifestation and a continuing moral support of the researches which built the newer knowledge of nutrition.

When in the summer of 1905 or 1906 a leading New York lady stood up in her open carriage on Fifth Avenue to shout across the street traffic, "I've got another!" she meant that she had secured a gift to support another "milk station," that is, a center for the distribution of properly prepared milk together with advice on the care and feeding of children. And scientific feeding of children spread as knowledge grew.

In 1931 on the occasion of an informal conference on the subject in Puerto Rico, the question was raised whether it would be wise to start such a child-feeding station as long as there was doubt that the milk supply would permit the project to be permanent. "Yes," said the Director of the School of Tropical Medicine, "let's begin now and continue as long as we can, for every day that the child does get the extra milk he puts calcium and vitamin A in his bank." The speaker was the late Dr. Earl B. McKinley, primarily a bacteriologist, yet appreciative of the newer knowledge of nutrition.

When during the First World War the American Red Cross sent a medical mission to Rumania, and when the fighting ceased a group of English physicians began medical relief work in Vienna, much evidence was found of the difference in development and health of children of the same race, place, and age, according to what it had been possible for their parents to feed them. Depletion of the deposits of vitamins A and D in the bodily "bank" and failure of sufficient calcium and riboflavin supplies during 1914–18 made differences which at that time may not have been translatable into these chemical terms but which were clearly correlated with whether the mothers had been able to get milk or eggs for their children or had been obliged to feed them on bread and water, or perhaps more literally on bread and turnips.

In a series of experimental investigations begun in the then new light of the human experiences of the First World War, as well as of the simultaneous advances in the chemical concepts of nutrition, nutritional-balance experiments with children and long-term feeding experiments with laboratory animals were carried on at the same time in the chemical laboratories of Columbia University, each of these two types of experiment being planned and interpreted in the light of the other; as also, of course, in the light of nutrition investigations of both types elsewhere. Furthermore, the laboratory-animal feeding researches at Columbia were planned in two ways: in certain series the experimental variables were the individual chemical factors such as calcium or vitamin A; while in other series the experimental variables were actual everyday articles of food, such as bread, milk, meat, or turnips.

This comprehensiveness of research plan is well justified by the results. The combined group of pediatricians and nutritionally minded biochemists who discussed the subject in the proceedings of the first White House conference on child development and health clearly recorded the modern professional view that, with investigators reasonably competent to select the right species to serve as a tool of nutritional research, the slight disadvantage of "species difference" may often be far outweighed by the much more complete experimental control permitted by the use of laboratory animals.

Laboratory-animal experimentation may serve much the same function in relation to human nutritional research that the "pilot plant" does to industrial research in development.

At the time this is written, several investigators are following up, in the medical assessment of nutritional condition in people of different ages, some of the relations of food to the onset of "senile" changes (or, conversely, to "the conservation of the characteristics of youth") which had been found in laboratory-animal experimentation between 1920 and 1940. With rats as instruments of research, the laboratory experimenter in nutrition—whose background is most often predominantly that of training and experience in the "exact" science of chemistry *plus* modern physiology's emphasis upon experimental controls and statistical interpretation—can run through research experiments upon complete life cycles thirty times as fast as would even conceivably be possible with human subjects.

So, if we could conceive of any arrangement (in our type of civilization) by which human beings in large enough numbers could be sufficiently controlled for full-life and successive-generation experiments, science would still be compelled to wait at least several centuries for the completion of researches directly upon human beings to cover the ground covered by experiments with rat families during the two decades of 1920 to 1940.

In one of the earliest of the Columbia experiments in which laboratory animals were used, young rats of the same litter were from weaning time fed upon different simple diets of staple foods as follows: bread alone; bread and meat; bread and apple; bread and turnip; bread and milk. The bread was always of the same kind, made from white flour with fixed (usual) proportions of table salt, yeast, and a vitaminless fat. In each diet which contained a second food, the latter furnished one fifth, and the white bread four fifths, of the total food calories. Of the diet thus quantitatively proportioned, each rat was allowed to eat the amount to which his voluntary activity and his rate of growth inclined him. Distilled water to drink was also always freely available. The results of these experiments were as follows. With bread alone there was no growth, and relatively short survival. With bread and meat there was good growth for a short time, then cessation of growth quickly followed by rapidly progressive weakness and loss of weight, death coming only a little later than when bread alone was fed. With bread and apple, there was no growth but somewhat longer survival. With bread and turnip (the diet of which we heard so many and such bitter complaints from Central Europe during and immediately after the First World War) there was only slow, decidedly subnormal, and irregular growth continuing as long as it was feasible to continue that particular experiment. With bread and milk there was growth at a normal rate to a normal adult size; but with four fifths of the food calories in the form of white bread and only one fifth in the form of milk there was no rearing of offspring. Successful reproduction and lactation with rearing of normal offspring became possible when the bread-and-milk diet was improved either by replacing white bread with whole wheat or by increasing the proportion of milk in the diet.

The experimental food mixture consisting of ground whole wheat and dried whole milk in the proportion of 4 to 1 by calories, or 5 to 1 by weight, with table salt and distilled water (Columbia laboratory diet No. 16), has been found adequate to support health, growth, reproduction, and successful lactation in our laboratory rats generation after generation. At the time this is written we have rat families thriving on this diet in the fifty-fourth generation. This is an exceptionally rigorous demonstration of the adequacy of a diet; and as we have made this the dietary starting point in many of our experimental series we have, for the convenience of readers, commonly called it Diet A in print.

While this *Diet A is adequate* in the sense that the word adequate is generally used and understood, it is not optimal. An increase of the proportion of milk in its wheat-and-milk mixture enables the animals nourished upon it to improve

their average records in growth, development, adult performance, and length of life; so the food mixture with higher proportion of milk, *Diet B, is better.*

The betterments of the average records at all stages of the
life cycle on Diet B as compared with Diet A have been established by large numbers of full-life and successive-generation experiments. The condition of nutritional well-being is
normal on Diet A and averages better on Diet B. Here the
nutritional improvability of the normal is clear-cut and undoubted in the objective records both of early and of later
life: growth and development are expedited, adult vitality
enhanced, senility deferred, and longevity increased, all in
the same individuals by the same dietary improvement.

These well-established research findings are undoubtedly
very significant in their human implications and applications
because the starting point of these laboratory feeding experiments (while simplified as to the number of articles of food
for greater accuracy of control and certainty of interpretation)
is quite analogous to the dietaries on which certainly a large
proportion, and probably a large majority, of people live. It
is adequate for the maintenance of passable health and the
perpetuation of the population, but is capable of improvement by scientific shifting of the quantitative proportions of
the foods which the dietary contains.

In this case the shift in proportions of natural foods meant
enrichment of the dietary in four of its chemical factors:
protein, calcium, riboflavin, and vitamin A value. When these
enrichments were tested *separately* as *additions* to the same
basal Diet A we did not find all aspects of the life history improved by increased intake of each separate nutrient factor,
as they all had been by the increased proportion of milk in Diet
B as compared with Diet A. The added protein increased the
rate of growth and the average adult size; but not the adult
vitality as judged by performance, nor did it increase the
average length of life. The added calcium did better the
record at every stage in the life history. The added riboflavin

has not been tested through a sufficient number of complete life-cycles for conclusive findings, but it seems to have a generally favorable influence throughout and to benefit the offspring. The added vitamin A, under the conditions of these experiments, did not increase the rate of growth but did increase the adult longevity.

Neither the rate of growth nor the bodily size attained is conclusive as an indication of well-being throughout the lifetime. Some pediatricians warn against too great an ambition to see one's child make exceptionally rapid growth, and are reinforced by the finding in the Agricultural Experiment Station at Cornell that a very rapid growth, induced by extremely rich diet given in unlimited amounts, may not always be best in its effects upon the life cycle as a whole.[1]

There is not necessarily any real conflict between the Cornell and the Columbia findings. If they appear divergent this may be simply because they deal with widely separate areas in what we now see to be a very broad field of investigation.

The starting point of the Cornell experiments was an extremely rich diet, such as may be approached in occasional cases of the forcing of farm animals for maximal gains in body weight; and possibly when an infant is fed without regard to economic considerations and with too great a desire to make a phenomenal record of growth at an early age. Such cases of undue forcing of growth by extreme richness of food, while they may occur in practice, do not seem likely to affect more than a small minority of the population.

The starting point of our experiments at Columbia was a dietary much more representative of the food supplies upon which the great majority of people must depend. The chief sources of food calories are the grain products, here represented by wheat, and the dietary is made adequate by the inclusion of an economical proportion of the so-called "protective" foods, here represented by milk.

[1] C. M. McCay, M. F. Crowell and L. A. Maynard, *Journal of Nutrition*, 1935, 10: 63–79.

Two findings which need not be taken up technically here should nevertheless be mentioned as greatly strengthening the grounds of our research procedure in studying the very important problem of the relation of food to bodily well-being at all stages of the life and the extent to which growth in itself affords an indication of one's bodily future.

The first of these is the conclusive establishment of the fact that the growth data of our experiments have such a symmetrical frequency distribution as to justify a high degree of confidence in their statistical interpretation.[2] The second is that faster or slower growth as an *individual* characteristic does not in itself influence the life expectation. That is, the *strictly individual* variations in growth and in longevity are independent, not interdependent.[3]

These two findings place our research program in a much stronger position for the solution of the further problems mentioned above as still under investigation.

While the improvements in health and longevity, to which the newer chemistry of nutrition is showing the way, involve nothing so dramatic biologically as a mutation or as some of the exploits of endocrinology, yet the nutritional improvement of life can have much more than a merely biological significance.

As the plan of the present book contemplates a small size and a wide scope, it is not practicable to assemble here the technical evidence underlying every statement. In this chapter particularly (and in varying degrees in some of the others) an adequate summary of the subject in hand requires the putting together of evidence of different kinds without stopping for detailed explanations. Different kinds of evidence may possess different degrees of precision and conclusiveness; but in all cases the basis is objective, not subjective. Everything herein contained is strictly scientific finding or interpre-

[2] Sherman and Campbell, 1934, *Proc. National Acad. Sci.*, 20: 413.
[3] *Ibid.*, 21: 235.

tation, as distinguished from anything speculative or into which wishful thinking might enter.

Through the objective findings of such investigations as that made by Corry Mann and that included in the Milk-in-Schools Scheme, to mention only two out of many, it is now to be regarded as a conclusively established scientific fact that nutritional improvement, even of a dietary deemed adequate according to current standards, can enhance the mental along with the physical development.

Without implying any disregard of other possible factors we may be entirely confident that wiser feeding is at least a major factor in the unquestionable finding that boys and girls now enter college both younger and taller than formerly. Moreover, we can now say with confidence that *the same* nutritional improvement can contribute both to the better mental development and to the better physique.

And if we turn from the college to the factory we find that with better feeding there is quicker learning of industrial processes and a higher efficiency more constantly and progressively maintained. The worker whose food habits are guided by the newer knowledge of nutrition not only loses less time but experiences less fatigue and less often injures either himself or his material.

Of course this does not mean that physique and mental development always run parallel. Either natural endowment or training or surroundings may affect the two quite differently. Nor is it correct to suppose that every dietary difference which influences the size of the body will correspondingly influence the mental capacity of the individual. In fact "feeding for size" may with some diets result in a larger size but no betterment even of physical stamina. It is only when an adequate understanding of the newer knowledge of nutrition is the guide that we may depend upon constructive improvement of the body's internal environment and the resultant improvement in the quality of life processes to induce superior growth and development both mental and physical, followed by a higher level of adult capacity, which will be maintained for a longer time.

This fact, taken with the new evidence that the internal environment is much more importantly flexible to the specific influences of different nutrients than ever previously supposed or yet generally appreciated, involves potentialities of exceedingly great importance.

For a food habit or nutrition policy which both expedites development and postpones old age makes possible careers of greater cultural achievement and social value.

In one of his annual reports as president of the Carnegie Institution, the late Dr. R. S. Woodward bemoaned the shortness of scientific or other creative careers, and remarked that a third of a professional or scientific man's years have usually passed by the time he has finished his formal schooling and entered his constructive life work; then probably another third will be spent in proving to himself and to others what he is able to do, before he will be entrusted with his highest responsibilities; and so, only the last third of his years remain in which to render his fullest service to the world.

A charting of the age incidence of major opportunities of several hundred presumably representative men in occupations of a scientific or related administrative or educational nature strikingly confirms Woodward's impression that the time of attaining (or being promoted to) such fullest opportunity is most frequently around the age of 50 years. Perhaps equally striking is the wide range of ages at which appreciable numbers of men have actually found their major opportunities.

Both these facts emphasize strongly the advantages to the individual and the gains to society which may confidently be anticipated from the earlier attainment and the longer retention of the full adult capacity and efficiency of individual persons (and of the constantly increasing proportion of people) who will have received the benefits of the newer chemistry of nutrition.

When we consider that, even under the present tradition with its bias toward delaying a man's promotion to his highest opportunity, many individuals do attain to their major opportunities much earlier than the average, it seems clear that early recognition will be accorded more frequently when through better nutrition young men actually do develop more promptly to full adult capacity, and when through spread of nutritional knowledge it becomes generally known that development of optimal adult capacity can be so expedited.

And the young man need not fear that the same spread of nutritional knowledge will act to delay his opportunity by enabling the oldsters to hold the higher jobs longer; for here the younger.man has two facts in his favor. In the first place, the younger one is, the more one *can* profit from the newer knowledge of nutrition. "Youth will be served" by this opportunity to better his actual life processes, if he will avail himself of it, to a greater extent than is possible for older men, other conditions being equal. In the second place, the knowledge that longer careers of full capacity have become possible will greatly facilitate the working together of younger and older men on essentially equal terms—the younger taking the more executive functions while the older serves in a more advisory or consulting capacity. Many and widely different kinds of work have been found to profit greatly by such collaboration to which the one brings fresh training and youthful energy; the other, mature judgment based on a rich store of experience.

Dr. Robert F. Griggs, chairman of the Division of Biology and Agriculture of the National Research Council, has written of the importance of the new scientific evidence which "shows that man properly nourished maintains a vigor in life never before thought possible." And many another executive, whether with or without the knowledge that the reason is largely nutritional, also has found that young men differ considerably in the rate at which they progress to full maturity of capacity and judgment; and that older men differ in still greater degree in the ages to which they retain their full capacities.

Undoubtedly it will become increasingly clear that the periods of full recognition and responsibility of the successive generations of men must be arranged not only "in series" but also to an increasing degree "in parallel," though (as suggested above) with some differentiation of function between the younger and the older men of similar rank. This is both for the good of the organization or the public which the men serve and for the fuller opportunity of each individual to enjoy every segment of a well-rounded life cycle.

In the adjustment of the overlapping careers of successive generations of men to their mutual satisfaction and to the best interests of human progress, doubtless the first essential is simply that the development and promotion of the younger men into ever higher responsibilities as early as earned shall never be obstructed by the presence of their elders in these jobs. The elders should stand ready to move out just as soon as their successors are fully ready to take over.

But this retirement from executive function and from competitive activity should not mean compulsory idleness. It should mean an increased degree of freedom in doing one's most mature work, regardless of whether the official status is that of a consulting expert in the organization, or semiretired, or retired.

In any such case the man who has relinquished competitive or managerial activity while still possessed of keen interest and good working capacity may garner the harvest of which Eliot after retirement from the presidency of Harvard spoke with such gratification. One's work in this period, he wrote, is especially appreciated by others, because they now see clearly that he is both experienced and disinterested.

The established fact that the current norms of the life history, or of the life process as a whole, can be nutritionally improved so that the period of the prime is reached somewhat sooner and then held much longer before the onset of senility appears carries in itself a greatly enhanced opportunity for the individual to enjoy a satisfying career of accomplishment and for the community to reap a fuller harvest from him. From both the individual and the community points of view, the higher level of cultural attainment which Dr. McLester mentions is no more than a simple interpretation of the fact that nutrition has shown how to increase the ratio of the years of highest attainment and productivity as compared both with the period of development which must always precede and the period of old age which normally follows.

And the longer lease of fully efficient life which the newer knowledge of nutrition offers may mean much, not only in terms of satisfactions to the individual and his family and friends, but also to his other fellow men. For as Merriam has pointed out in his essay entitled, "Are the Days of Creation Ended?" [4] the direction of human evolution is now largely social, and society is a continuing organism interested in its own future. What promises to affect this future should influence our decisions from day to day and will do so more effectively with the growth of the scientific spirit which expects progress and works for it; but meanwhile the shortness of individual lives tends to set a limit to the actual use by man of the knowledge which he has accumulated and the institutions which he has built and developed. Hence the longer term of fully efficient years may be of far-reaching significance to human progress in affording fuller opportunity for the use of man's ever-growing body of knowledge.

Considerably more is here involved than merely a prolongation of technical or creative competence. The wisdom of our human family tells us that late in a well-spent life there comes a time: When old experience doth attain to something of prophetic vein.

[4] J. C. Merriam, *The Living Past* (Scribner's), 1931.

Dean Inge's quotation to the same effect is that longevity may be felt not so much in the weakening of the body as in the strengthening of the soul.

Certainly among the most far-reaching contributions of science to civilization is this, that the new chemistry of nutrition points the way to such improvement of the life process as to permit full realization of the potentialities both of youth and of maturity.

Shakespeare's intuitive insight could register, even in his prescientific time, that old age though frosty can also be kindly as the result of good use of the years of youthful stamina. What a thing it would be for the world if another seer like Shakespeare could now arise to tell in the universally understandable and memorable terms of human character, how the Will can implement itself with scientific knowledge for the heightening of the effectiveness of life.

CHAPTER XIII

NUTRITIONAL GUIDANCE FOR "THE BACK-WARD ART OF SPENDING MONEY"

IN AN essay entitled "The Backward Art of Spending Money" written some decades ago, our distinguished economist Wesley Mitchell developed a very significant thought, entirely valid so far as it went, but to which a sequel should now be added.

In general, he argued, people of our type of civilization make money more satisfactorily (in normal times, of course) than they spend it. This is because our gainful enterprises—in agriculture or industry, for example—are conducted under the guidance of our exact sciences such as physics and chemistry. When, however, we seek to convert our financial gains into satisfactory living, our efficiency proves to be relatively poor; and this is presumably because, whereas exact science guides our making of money, the spending of it to secure a satisfying life has only the guidance of such inexact sciences as sociology and psychology.

That the art of spending money has for some reason remained rudimentary, most thoughtful people will admit. Also most of them probably consider that the spending of public money is nearly if not quite as backward an art as the spending of individual or family income.

But something has happened which no one foresaw when that essay on the backward art of spending money was written. Nutrition, formally classified as a branch or aspect of physiology, and practically treated as an essentially autonomous science, has actually developed as a branch of the exact science of chemistry. In the light it now can bring to the wise use of that large part of the family income which must usually go for food, the newly exact science of nutrition furnishes that guidance the lack of which hitherto has kept the art of spending in its backward state.

The science of nutrition as it is today, with the potentialities which its recently discovered facts and especially its new fundamental concepts confer and with its ever-growing exactness and depth of insight, does bring to the spending of money the definiteness of scientific guidance which Mitchell so clearly saw that it lacked and needed, while no one until recently could tell whence it would come.

Wages and prices have undergone much recent change but a few figures may assist our sense of proportion. If three meals a day at 20 cents each were the per capita allowance of our population of about 130,000,000 people, the total food bill of consumers in the United States would be about $28,470,-000,000. Undoubtedly the science of nutrition if properly used could very greatly improve the returns from the expenditure for food whatever its amount. Certainly, too, the efficient use of this money is of great importance to welfare, for it constitutes a large fraction of the usual or expected total income.

Even more important, however, is the fact that health, efficiency, and satisfaction in life depend upon the investment for food more than for almost anything else that money can buy.

It has often been said, and will doubtless long continue to be true, that the problem how to bring the newer knowledge of nutrition most fully into the service of the largest number of people is both economic and educational. It is important that both these parts of the problem be clearly and frankly recognized. It is also important to emphasize that while both economic and educational efforts are needed, we are never justified in delaying one while waiting for the other.

There is ample objective evidence that food habits can be improved by education in nutrition and food values, whatever the level of income and purchasing power.

And it has also been clearly shown that even without further education the great majority of low-income families do provide themselves with nutritionally better dietaries as soon as their purchasing power is increased.

Each of us who can do anything for nutrition education should do it at once and continually, and with confidence in its value, however keenly we may feel that there should also be some way found, some means provided, to improve the purchasing power of the low-income families of our nation. Similarly each one who can do anything for direct economic betterment may do it with full expectation that it will make for improved nutrition and health, however true it may be that more or better nutrition education is also needed.

And all of us may constantly strive for still fuller recognition of nutrition in all publicly supported and officially administered economic projects dealing with food, such as crop-production programs, food-stamp plans, and school lunches.

Usually and mainly, improvement in the art of spending food money begins at home. Such a beginning in practice may be stimulated by the teaching of nutrition and food values. As a rule there must be a tangible consumer demand for more of "the right kinds of food" before farmers and processors will change their production programs, and usually, though not always, before the Government will advise them to do so. If Government takes the lead in advising a shift in food-crop production for the sake of better nutrition of the people, the advice will in the long run be effective only to the extent that it is supported by consumer demand in the market.

Sanitary improvements may often be made effective by Government action without the demand or even the knowledge of the people generally; but only in limited degree can this be true of nutritional improvements. Mainly, these must come about through the initiative or at least with the active support of consumer demand for the right kinds of food.

It is true, as already suggested, that the problem of providing a nutritionally satisfactory food supply is harder the lower the income, and will to some extent ameliorate itself if real income rises. However, it is also true that according to differences in choice of food, some people get dietaries nutritionally good, others only fair, and still others only poor, when all are spending at the same level in the same locality.

How, then, may nutritional guidance be given to food consumers? To answer this question as thoroughly as present knowledge permits would require not only the whole of this chapter but the whole of a book bigger than this one. Such thoroughness is usually sought only by those preparing to be professional dietetians or teachers of nutrition and dietetics. For such degrees of dietary guidance as most people desire, probably the most practical plan is to divide foods into from five to twelve groups and then indicate the desirable prominence of each in one of four ways: (1) what proportion of the total food calories is taken from each food group; (2) what proportion of the food money is spent for each type of food; (3) what actual amount of each kind of food is eaten per week or per year; or (4) how often each kind of food is served.

Of these four ways of gauging or expressing the prominence or quantitative place of a given kind or type in the dietary, each person in his or her practice or teaching may use the one which best fits the given conditions. Some people, perhaps, may use them all, each for the cases to which it is best adapted or in which it seems most practicable and instructive.

First Plan.—A guide originally suggested by Miss Lucy Gillett of the Community Service Society (here slightly adapted) is to think of the *food money as divided into five more or less equal parts* to be spent for five groups of food as follows:

One fifth, more or less, for fruits and vegetables;
One fifth, or more, for milk and cheese, cream and ice cream;
One fifth, or less, for meats and fish, poultry, and eggs;
One fifth, more or less, for breadstuffs and cereals;
One fifth, or less, for fats, sweets, and food adjuncts, or "extra" and "miscellaneous" items.

Second Plan.—The New York City Board of Health has set

up the following eight groups, not covering all foods, but recommending that *each of these eight groups be represented in the food of every day:* (1) Milk and its products; (2) Bread-stuffs and cereals; (3) Meats, fish, poultry, eggs, or meat substitute; (4) Citrus fruits; (5) Other fruits; (6) Salad vegetables; (7) Potatoes, sweetpotatoes, yams, plantains; (8) Other cooked vegetables.

Third and Fourth Plans.—The Federal Security Agency makes similar use of eight slightly different groups with some recommendations on the frequency of servings from each, as follows: "Every day, eat this way." (1) Milk and milk products: "at least a pint for everyone—more for children—or cheese or evaporated or dried milk." (2) Oranges, tomatoes, grapefruit, or raw cabbage or salad greens—"at least one of these." (3) Green or yellow vegetables: "one big helping or more—some raw, some cooked." (4) Other vegetables and fruits. (5) Bread and cereal: "Whole grain products or enriched white bread and flour." (6) "Meat, poultry, or fish; dried beans, peas, or nuts occasionally." (7) "Eggs: at least 3 or 4 a week" . . . (8) "Butter and other spreads: vitamin-rich fats, peanut butter and similar spreads."

The Bureau of Home Economics, United States Department of Agriculture, makes use of the twelve food groups enumerated in an earlier chapter and is publishing a series of folders, circulars, and bulletins with precise examples of the detailed working-out of family food supplies and meal plans at different economic levels.

By way of comment upon the above plans, it may be said that the first was drawn up with the people of lowest incomes most prominently in mind. Hence its suggestion of as much money for breadstuffs and cereals as for fruits and vegetables. Its wording, however, gives implied approval for the shift of some of this bread money to increase the purchase of fruits, vegetables, or milk whenever this can be done without reducing the total calories below the family's need.

The first and second plans bracket meat and eggs while the

third and fourth plans separate them. For the present purpose it seems a more effective form of nutritional guidance in food economy to leave them bracketed as in the first and second plans. To treat eggs as an alternative to meat is nutritionally well justified and is good modern practice from the viewpoint of meal planning. On as many days as the family pleases, then, let eggs take the place of meat. The nutritive value of the dietary will almost certainly be fully maintained and usually there will be some saving in cost which might well be invested in extra fruit, or perhaps in ice cream. On the other hand, the extra cookery and extra expense of having meat and eggs on the same day, three or four times a week, as the third plan implies, do not in the opinion of the present writer bring any adequately compensating advantage.

The second plan if desired may be simplified by combining its seventh and eighth food groups.

Full quantitative food plans for different levels of expenditure are distributed by the Bureau of Home Economics, United States Department of Agriculture, Washington, D.C. Or for very explicit meal plans and unusually interesting and informative text see Mary S. Rose's *Feeding the Family* (Macmillan, 1940).

Inasmuch as many expert plans still concede more to expensive high-protein traditions than the writer believes to be the best nutritional guidance to the wise spending of the food money, he here offers again the twin suggestions that seem to have been found useful for nutritional "check-up" whatever the level of expenditure: (1) Allot at least as much for fruits and vegetables as for meats and fish; (2) Allot at least as much for milk in all forms (including cheese, cream, and ice cream) as for flesh, fish, and fowl.

Now that we know our daily food to have a much greater influence upon our life-long efficiency than we ever previously supposed, we should not hesitate to invest a larger part of our individual and family incomes in food; and to advise others to do so.

It is appropriate to restudy, both the place of each type of food in the food budget, and the place of the cost of food in the total cost of living. Are we too complacently tradition-bound in our habit of investing too high a proportion of our food money in the so-called main dish of the dinner, forgetting that some of the foods traditionally placed farther from the center of the menu are the ones which yield us the greatest return of nutritive value for money cost?

One may well read again Table 7 in Chapter IX and also study Table 8 herewith. The data in Table 8 are based on those in Table 7; but again adapted for convenient coördination with the present discussion. Now that more has been explained of the importance to health and well-being of liberal margins of intake above "actual" or directly demonstrable "requirements" in the cases of calcium, riboflavin, and vitamins A and C, it will be of interest to learn from these tables what are the most profitable types of food in which to invest from this point of view. Of course, too, we always need ample supplies (although not large surplus margins) of calories and protein of which the breadstuffs and other grain products, at least in their staple forms, are outstandingly economical sources.

Thus, in the cases shown in Table 8, an expenditure of only 18 percent of the food money for breadstuffs and other grain products brought in return 31 percent of the total food calories and 28 percent of the total food protein, but less than *pro rata* shares of calcium, riboflavin, and vitamins A and C.

Milk, however, brings much more than its *pro rata* shares of calcium, riboflavin and vitamin A; and also a full quota of protein of such nature as very effectually to supplement the proteins of the grains. Here our most modern knowledge of nutrition and food values reinforces the time-tested advice that the dietary should be built around bread and milk: the bread for its cheapness so far as it goes; the milk as most effectively balancing the bread. More vitamin C than the bread and milk provide is, however, desirable; and additions to the

Collective Food Group	Proportion of Money Spent (Percent)	Proportion of Total from Each Collective Group						
		Calories (Percent)	Protein (Percent)	Calcium (Percent)	Phosphorus (Percent)	Vitamin C (Percent)	Riboflavin (Percent)	Vitamin A value (Percent)
(1) Breadstuffs and other grain products	18.1	31.3	28.2	11.2	20.2	1.4	5.5	4.1
(2) Milk and its products	17.0	16.0	17.7	64.9	29.0	5.3	35.6	22.4
Sum of Groups (1) and (2)	35.1	47.3	45.9	76.1	49.2	6.7	41.1	26.5
(3) Fruits and vegetables	19.9	14.1	12.2	15.7	20.0	91.6	20.7	57.7
Sum of Groups (1), (2) and (3)	55.0	61.4	58.1	91.8	69.2	98.3	61.8	84.2
Sum of Groups (2) and (3)	36.9	30.1	29.9	80.6	49.0	96.9	56.3	80.1
(4) Meats, fish, poultry and eggs	35.2	26.8	40.6	6.8	28.9	0.8	37.3	15.5
(5) Lean meats, fish, poultry and eggs	29.6	14.1	39.3	6.4	28.1	0.8	36.6	15.5

riboflavin and vitamin A values of the dietary may also be good investments.

Here the fruits and vegetables are of special efficacy and interest. In the presumably representative dietaries summarized in Tables 7 and 8, an expenditure of one fifth of the food money for fruits and vegetables brought over four times its *pro rata* of vitamin C, nearly three times its quota of vitamin A value, and its full share of riboflavin, as well as noteworthy returns in calories, protein, calcium, and phosphorus.

As may be seen from the fifth row of figures in Table 8, the three groups of foods just mentioned, when taken together, constituted a better investment in food values with respect to all seven of the factors here considered, than did the remainder of the dietary when considered as a whole. It is, of course, also true that if one sought enrichment of the dietary in calories alone, sugar and fat would appear as economical sources; or if interested in protein enrichment alone, or in protein and riboflavin without regard to calcium and vitamin A, the lean-meat type of food might appear outstanding; but these latter viewpoints would rarely be scientifically sound. At most times and places in the United States, the emphasis of nutritional guidance in the spending of money for food will be: (1) give a higher place in the food budget to fruits, vegetables, and milk; (2) give breadstuffs and staple cereals a higher or lower place according to need for high calories at low cost; and (3) let all white breadstuffs be enriched.

As to making the first of these suggestions more explicit, the writer recommends that the aim be to take at least half the food calories *as,* or spend at least half the food money *for,* fruits, vegetables, and milk in its various forms, including cheese, cream, and ice cream.

And in the *budget of the family's cost of living* might we not well allot to food a still larger part of our total expenditures in view of the fact that the influence of a well-chosen dietary is now known to be so much more constructive and far-reaching than could be known or scientifically suspected by any previous generation?

Of course, it is all too true that a suggestion to shift money to the food budget from any other line of expenditure is apt to go counter to some prejudice. And these prejudices may have entrenched themselves on high grounds: among the commonest are that to reduce the budget allowance for *that* would be to impair the American standard of living, *or* to rob the poor of some part of their few pleasures.

Actually, investment in good nutritional status will usually do more than any other investment of income to improve the true level of living, both because of its inherent importance and because it is so fundamental to increased earning power. For the same reasons and because nutrition so greatly influences the children's happiness, development, and prospects of successful and satisfying lives, good investments in food bring more pleasure in the long view to all members of the family than any other use that could be made of the marginal money.

As people come to understand how much right feeding can mean to their own and their children's lives, they will be glad to add to the food money *a part* of what has been going for "Amusements other than automobile," and "Adornments other than clothing."

Proposals to spend public money for the attainment of the higher levels of nutritional well-being which we now know to be dependent upon "enough of the *right kinds* of food" may also meet a prejudiced opposition. One form of prejudice which is very pervasive in all such connections as this is the resentment of businessmen against anything which they consider to affect their respective business interests unfavorably or less favorably than it affects someone else; and obviously a nutritionally guided improvement in food consumption will increase the market for some foods more than for others. Against any such use of public money, prejudice can (and often does) intrench itself on grounds of an interpretation of business ethics which opposes discrimination against any legitimate commodity, or on the ground that scientific guidance to a better use of food is "mere theory" evolved by people ignorant of or prejudiced against the realities of economics. The combination of tradition and sales effort through which consumers can be induced to buy what the interested food industry wishes to sell in at least as large quantity as ever before is much bolstered by impressive-sounding use of the word economics to keep consumers in a traditional attitude toward the place of the given food in the dietary.

Sometimes as further entrenchment of the "vested interest" point of view and sometimes on grounds of political principle, "subsidized consumption" is opposed as something that Government should not engage in because it is too paternalistic or tends to pauperize the people. Notwithstanding an original predilection toward this sort of political principle, the present writer believes that whatever validity it may have had is now far outweighed by the changed aspect of American opportunity and the new knowledge of the very great influence of enough of the right kind of food upon individual development and public health. So long as the nation offered all its people the opportunities of a freely open frontier, other forms of paternalism and subsidy were properly held on the defensive. But with the disappearance of the frontier the American ideal of fair chances for all came to need other implementation and it has been held not-too-paternalistic to offer greatly advanced educational opportunity at public expense, and to support an ever-expanding public-health service. While previously it was the opinion of some individuals it is now objective knowledge that for a large percentage of our population the educational opportunity cannot be effective without some provision for adequate nutrition. A reasonably good nutritional status is just as necessary as is a public school to give the child the fair start that the American ideal demands.

And whatever may be the future interpretation of the four freedoms which present-day America is said to stand for, it seems certain that with present knowledge properly diffused it will be recognized as a public responsibility that no child shall be allowed to remain chronically in want of *the food that he needs for his health.* For it is now clear, to anyone who will study the evidence, that nutrition has greater constructive potentiality than science had foreseen; and that even in the everyday choice of food we are dealing with values which are above price, for the health and efficiency, duration and dignity, of human life.

CHAPTER XIV

NUTRITION POLICY

ONE may enter upon consideration of nutrition policy from either the personal or the public angle of approach. Let us consider each of these and something of their interrelations.

The new science of nutrition enables those who choose to use it to "engineer" their own life processes in higher degree than has been possible before. Each person in deciding to accept the guidance of scientific knowledge in his daily food habits is adopting a nutrition policy for himself. And the grownups of each family may at the same time shape their own and inaugurate their children's nutrition policy, knowing that the second generation (with whom the benefit thus begins earlier in life) may be expected to profit even more than did their parents.

Knowing as much as we do of the great influence of nutrition upon the well-being of a family from generation to generation and upon the prospects of success and satisfaction for each member in whatever he or she undertakes, it is now quite clear that a family should as logically have a nutritional as a financial or an educational policy.

In a certain very true sense it might be said that the nutrition policy of a democratic nation is essentially the sum of the nutrition policies of its people as individuals and families, plus whatever of governmental activity the people as voters may decide to add. Probably to most readers, however, the word policy will tend rather to connote the public approach, the pursuit of desired objectives through the governmental machinery of the nation or other political unit.

Whatever the future forms of political organization may be,

it will doubtless continue to be true that the people of a given country or region, with a feeling of unity and with freedom or autonomy of action as a nation or state, will normally desire to advance both the quantity and the quality of the life of their kind. Their policy will contemplate the maintenance certainly and probably the increase of their population; and at the same time will seek progress in the well-being and cultural attainment of the individuals and families of whom the population is composed. In this present middle segment of the twentieth century we have most urgent need to accentuate the functioning of our new knowledge of nutrition that we may gain in health, strength, and efficiency to preserve our civilization and to provide for its humane and orderly advance.

So obviously is health important to the well-being of the people at all times and to the maintenance of the life of the nation in times of war that public health has long been recognized as a governmental responsibility.

Only now, however, are we coming to understand how predominant is nutrition among the factors upon which health depends. With this new knowledge it becomes as clearly "a responsibility of the public exchequer" to see that the people have nutritionally good food supplies as to see that they have schools, roads, water supplies, and protection from epidemics of infectious disease.

The present situation has developed somewhat as follows. Scientific research revealed: (a) the existence in certain foods of previously unimagined substances capable of curing and preventing what had been baffling diseases; (b) the fact that good diets in preventing these newly explained deficiency conditions may also be "protective" in the further sense of diminishing the incidence, or the severity, or the duration of some of the infectious diseases; and (c) that scientifically guided nutrition can be not only curative and preventive but actually constructive in building already normal health and efficiency to higher levels. And we both want and need as high a level of health and efficiency as we know how to attain.

Our Federal Department of Agriculture in its official Year-book for 1939 (p. 33) suggested nutrition policy without using the phrase but "as an attempt to answer three questions: What do we need in order to be well nourished? Do we get what we need? If not, how can we get it?"

At the National Nutrition Conference called by the President of the United States in 1941, representatives of a wide range of governmental departments and agencies strongly recommended national nutrition policy in such senses as are implied in attempts to answer these three questions not only on paper but in practice. This policy is extended beyond the United States in the carrying out of the Lease-Lend law and in coöperation with such other nations as contribute to the preservation and advancement of the civilization in which we believe.

How, then, can we best define in workable terms: (1) what food supplies we need in order to be well nourished; (2) the extent to which we are getting such food supplies now; and (3) how we individually and nationally, as families and as a human family, can have nutritionally satisfactory food supplies?

Each of the preceding chapters, while primarily in the form of a self-contained story of scientific advance, has also in some degree contributed toward the answering of these questions. To summarize fully the contributions of those chapters would make this one too long, so we shall here assume that the reader has them reasonably in mind, and shall refer only to the particular points which are most directly involved in the attempt to focus the light of scientific knowledge upon the practical objectives of nutritionally good food supplies. We put this in the plural because such food supplies need not all be alike. As McCollum has wisely said, we may eat what we like while eating what we should.

Scientific principles are now sufficiently developed to be of very real aid in the practical problem how best to use food for the promotion of individual, family, and public well-being.

For the sake of clarity one must sometimes repeat—what to the scientifically minded may seem a truism—that while in practice we nourish ourselves with food commodities, yet strictly speaking the absolute needs are the specific *nutrients* or *nutritional factors*. The National Research Council "yardstick" provides quantitative allowances for ten of these factors, in the belief that a dietary which is composed of reasonably natural foods and which meets the recommendations on these ten points may normally be trusted to contain adequate amounts of the other essentials.

The usual present-day practice is therefore to judge the adequacy of food supplies by reference to this yardstick. But the actual facts of nutritional need are not quite so rigidly and mechanically defined; for the body is more than a machine. For example, carbohydrates and fats while serving as fuel have also a protein-sparing action in the body, so that when the total calories of the dietary are liberal the body then has an even greater margin of safety in regard to protein than would appear from a study of the figures for protein alone. Another example is found in the fact that while each vitamin has one or more specific functions, yet among some of the vitamins of the B group there are interrelationships such as, for example, that a liberal amount of riboflavin in the food improves the body's nutritional insurance as regards both riboflavin and thiamine.

Chemically different as are these two examples of the preceding paragraph, they illustrate a principle of far-reaching importance, namely, that the nutritional processes possess a certain flexibility of scientific character beyond what can be correctly represented by machine-model analogies or simple bookkeeping calculations alone. Hence considerable flexibility of food supply is consistent with excellence of nutrition. It may be just as well nutritionally, and perhaps better for economic and psychological reasons, not to urge any one dietary pattern too strongly upon all people. The chemical requirements of human nutrition are doubtless much the same the world over, and in these terms we should not expect other peoples to do their best on less than we recognize as optimal for ourselves; yet the optimal supply of specific nutrients or chemical factors may be attained through fairly widely different supplies of food commodities.

Food habits and traditions which afford social satisfaction can be reasonably retained while at the same time we make use of the guidance of the new science of nutrition. For as stated by the United States Department of Agriculture in its official Yearbook for 1939 (p. 7): "This science of nutrition does not wipe out habits and traditions. It supplements them, corrects them, and shows how to use them intelligently. It offers a sound foundation for the food production and distribution of the future."

This chapter is written in the spirit of the preceding paragraph, and at a time when problems of nutrition policy are more important than ever before. For we see more clearly than was previously possible the importance of nutritionally good food supplies to human welfare, and at the same time we face a long period in which world conditions may accentuate the difficulty of ensuring enough of the right kinds of food for all of the people.

What Do We Need in Order to Be Well Nourished?

If our concern with this question were confined to the purposes of scientific discussion in itself, we could answer that we need adequate amounts and well-balanced proportions of all the chemical factors which we depend upon food to furnish.

Can we make an equally concise statement of nutritional need in terms of the articles of food that nature and agriculture produce and that are normally marketed and consumed?

Lord Astor's simplification, that nutritional need is "not only enough food" but also that this shall include "enough of the *right kinds* of food," has been helpful in awakening nutrition consciousness, but needs to be made more explicit before it can function far in the actual guidance of practical policy.

A grouping still too often heard is that which distinguishes merely between fuel foods (or high calorie foods) and "the proteins." This, of course, is not only inadequate but very seriously misleading. If a first division of foods into two main types is wanted, we may start in either of two ways: On the first plan we may distinguish the foods chiefly significant for their fuel values from the foods chiefly significant as sources of proteins, vitamins, and mineral elements. Or, on the second plan, we may distinguish foods which chiefly furnish calories and protein from foods chiefly significant for their mineral and vitamin values.

The latter of these two first-divisions has the great advantage of bringing the newer knowledge more clearly and immediately into due prominence in even the very simplest of distinctions among foods. This has been the outstanding merit of McCollum's contribution to nutrition policy.

Stiebeling and others of the United States Department of Agriculture, working largely from the viewpoint taught by McCollum yet seeking to carry the grouping of foods into enough different categories for the practical purposes of meal planning and of explicit recommendations as to family food supplies, find useful a division of foods into the following twelve groups: (1) Milk with its products other than butter; (2) potatoes and sweetpotatoes; (3) dry mature beans, peas, and nuts; (4) tomatoes and citrus fruits; (5) leafy, green, and yellow vegetables; (6) other vegetables and fruits; (7) eggs; (8) meat, poultry, and fish; (9) breadstuffs and cereals; (10) butter; (11) other fats; (12) sugars. Stiebeling and Clark in their article entitled "Planning for Good Nutrition" (pages 321–40, of the United States Department of Agriculture Yearbook for 1939) show how well this grouping works for the purposes for which it was designed. It aids the home-maker who wishes to feed her family nutritionally good meals while at the same time catering to their habits and preferences.

"Are We Getting What We Need in Order to Be Well Nourished?"

As a previous chapter has been devoted to so nearly this same question, and as many of its aspects are implicit also in parts of several other preceding chapters, a brief answer will suffice here.

All well-informed opinion is fully agreed that in most parts of the world, including our own, large proportions of the people are not getting the food that they need in order to be well nourished.

Any apparent difference of competent opinion on this point may be attributed either (1) to the fact that recent growth of knowledge in this field has been so fast and has reached so deeply into fundamental concepts that some people who might be expected to be acquainted with nutrition have not kept up to date, or (2) that differences not only of knowledge but also of temperament enter into today's judgments as to how large should be the "margin of safety" or the "insurance" in the nutritional allowances which we should recommend in the light of our new knowledge.

For reasons already explained, even severe cases of malnutrition are very incompletely recorded as yet, while the very much larger numbers of cases less clearly marked are apt to go entirely unrecognized. Also, the more delicate methods of diagnosis now coming into use agree with the statistical studies of family food consumption in revealing that a large percentage of our people have not regularly received all that is needed in order to be well nourished.

There certainly is no scientifically sound justification for complacency in the present situation. But our new knowledge of nutrition is constructive as well as corrective; so that while there is more repair work needed than we previously knew or supposed, the same new knowledge which shows this fact serves also to point the way to advances beyond what had been anticipated. Hence the well-trained nutritionist or nutritionally minded physician can be not only a repair man and preventer of accidents but also an architect of the higher health.

This higher health is not to be thought of as an ideal to be attempted sometime in the future: It is an immediate problem whether our world neighbors as well as our own fellow citizens have food supplies which will support the development and maintenance of their full potentialities for health, strength, stamina, and endurance.

In this country and Great Britain probably only the poorest of the poor are unable to find the food calories they need. There are large reserve stores of wheat in both these countries. In Russia and China, however, there are large areas in very great need of bread or its equivalent.

The general situation is very similar for food protein as for food calories. In Great Britain and perhaps in the United States there is some economizing; but in neither is there such a shortage as to endanger nutritional well-being, while there doubtless is such shortage of food protein in parts of China, and probably also in parts of Russia. In both these latter countries, it will be remembered, the invaders have, over large areas, consumed, destroyed, or carried away not only the annual crop but also the farm livestock. Much time will doubtless be required for rehabilitation.

Present evidence indicates that the most frequent shortages of specific nutrients in American dietaries are of calcium, vitamins A and C, and riboflavin. It would be premature to try to indicate the order of frequency of these four shortages. Perhaps thiamin (in the United States generally) and niacin (in some regions) have been among the most frequent deficiencies in recent decades, but these dangers are, we may hope, now being importantly alleviated by the enrichment of most of the white flour and bread used in the United States.

In Great Britain the nutritional shortages of which there is most danger are probably the same as in this country. Correspondingly the foods which we should send in relatively greatest abundance are cheese and dried or evaporated milk, and citrus fruits or canned tomatoes or their concentrated juices. Meat should also be sent liberally to our British and other Allies until the per capita meat supplies of the United Nations are brought into reasonable relation.

In parts of China and Russia there is, as briefly noted above, great need of shares of our grain crops when these can be shipped; and there are also severe shortages of fats, meats, eggs, and milk, because of the extent to which herds and flocks have been devoured by the invaders. The replacement of farm livestock will, of course, be an even longer process than the restoration of normal cropping of the fields.

How Provide the Food That Is Needed in Order That All Shall Be Well Nourished?

STIMULATION OF AWARENESS OF THE PUBLIC IMPORTANCE OF NUTRITIONAL IMPROVEMENT

In 1931, Sir Frederick Gowland Hopkins pointed out in the first issue of the international quarterly *Nutrition Abstracts and Reviews* that right nutrition plays a larger part in the capacities and status of men and of nations than had been thought. Writing from the viewpoint of his broad knowledge of both chemistry and biology he emphasized the fact that Nurture can assist Nature (hereditary endowment) to a larger extent than had been realized or anticipated. He explained the mistake of supposing that a community which survives must be getting what it needs from its environment and that the limitations of its accomplishment can therefore be considered hereditary. Rather, it is more probable that few if any communities have ever had fully optimal food supplies for all their people; that each community which has survived has merely come into a sort of equilibrium with its food supply; so that with a nutritionally superior food supply such as science is now learning to provide, the level of capacity, accomplishment, and status of the community can be raised. And in another carefully considered statement written five years later he reaffirmed this interpretation and characterized nutrition as "a national problem of unsurpassed importance."

The question as to what proportion of the British people have a food supply sufficient for the development and maintenance of the standard of health which should be the birthright of their racial inheritance was actively debated in Parliament and anxiously studied by scientific workers; and from Downing Street there was sent to the British colonies and dominions throughout the world a state paper entitled Nutrition Policy which proposed that throughout the commonwealth of British Nations, with their dominions, colonies, and mandates, the making of tariffs, the setting of quotas, and other such actions shall be guided by a definite policy of bringing into human consumption the kinds and amounts of foods most conducive to optimal health.

In this country the official report of our Federal Secretary of Agriculture stated that the goal of agricultural research and adjustment is the optimal nutrition of the people; and a 1937 news item informed us that the then new schedule of allowances to farmers for coöperation in the Soil Conservation Program was designed to encourage dairy farming and vegetable and fruit growing.

To safeguard the health of the people has long been recognized as a function of government, and now that nutrition has shown an hitherto unexpectedly high potentiality for improvement of health, efficiency, and welfare, the concept of what is appropriate governmental activity in this field is fundamentally altered, and nutrition policy becomes not only a legitimate but an urgent governmental concern and responsibility.

This was recognized by the League of Nations which appointed, under the chairmanship of Lord Astor, a "Mixed Committee on the Relation of Nutrition to Health, Agriculture, and Economic Policy," whose Final Report was published in 1937.[1]

In that report it was stated (p. 34) that as contrasted with the previous "groping" of men for a better and more abundant life, "what is now required is the conscious direction" of the tendency towards better nutrition and that "Such direction constitutes policy." The report then points out that nutrition policy should be directed toward mutually dependent aims: (1) bringing the foods which modern science has shown to be needed for high health and physical development within the reach of all sections of the community; and (2) concerning itself with the necessity for facilitating the adaptation of agriculture to changes in demand and for increasing supply as demand expands, recognizing that adjustments are always required whenever social progress occurs.

And the report concludes its general summary with the words (p. 53): "The malnutrition which exists in all countries is at once a challenge and an opportunity; a challenge to men's consciences and an opportunity to eradicate a social evil by methods which will increase economic prosperity."

Soon after the distribution of that report, the British Medical Association took the initiative in organizing a conference on nutrition policy for Great Britain, which met in London on April 27–29, 1939. It urged upon the British Government the formulation of a long-term food policy in which the requirements of health, agriculture, and industry shall be considered in mutual relation. It also declared its conviction that measures to secure the more ready availability to all sections of the community of foodstuffs which are held to be desirable on nutritional grounds should be accompanied by an educational campaign to encourage their increased consumption. Consistently with this general view and even before the emer-

[1] *Nutrition: Final Report of the Mixed Committee of the League of Nations on the Relation of Nutrition to Health, Agriculture, and Economic Policy.* (Printed in English and distributed in the United States of America by the Columbia University Press, New York City.)

gency measures of the Second World War, the British Government subsidized the growing of increased crops of those foods which, under the guidance of the newer knowledge, were held to bring most benefit to the consuming public. It established marketing boards, operated at government expense, to ensure for such foods as milk, eggs, fruits, and vegetables the lowest possible price to consumers with an adequate return to the producers. Also a school lunch of milk is given to all school children: at cost, or for what they can pay, and at public expense to those who can pay nothing.

According to the United States Department of Agriculture (Yearbook for 1939, pages 40–41), "It is true that we are only at the beginning of knowledge about human nutrition, but it is also true that enough is now known to give better health, greater vigor, and longer, more useful lives to immense numbers of people if the knowledge could be generally applied."

This Federal Department also states (*ibid.*, p. 41): "There is nothing mysterious about the practical application of modern knowledge of nutrition. Leaving out all the technical details, it says simply that the majority of people need to get more milk and milk products, eggs, and certain fruits and vegetables than they now get."

Vice President Wallace in his address to the National Nutrition Conference of May, 1941, suggested a nutrition policy for this country as follows:

It was found that certain types of disease like pellagra and beri-beri could be wiped out 100 per cent by proper food. We have had fairly complete knowledge about these dietary deficiency diseases for at least ten years but there are still too many people dying from them. Therefore I propose as goal number 1 of the National Nutrition Conference the complete wiping out of deaths caused by dietary deficiency. We do not have yellow fever any more in the United States. Neither should we have pellagra.

As goal number 2, I would propose a great reduction in those diseases such as tuberculosis toward which insufficient food predisposes. . . . Undoubtedly, we can reduce the death rate from these diseases by many hundreds of thousands by adequate food.

The third goal which I would suggest for this Conference excites me in some ways even more than goals number 1 and number 2. This goal is to make sure that everyone in the United States has in his diet enough energy, enough bone-, blood-, and muscle-building food, enough vitamins, to give that feeling of "health-plus." We do not want merely to wipe out pellagra, rickets, and scurvy and to reduce death losses from tuberculosis, but we want to make sure that our millions are so fed that their teeth are good, their digestive systems healthy, their resistance to premature old age enhanced through strong bodies and alert minds.

With food so critical a factor as it has recently been found to be, and with the urgency of the problem of optimal nutrition and of nutritional rehabilitation, we shall do well to emphasize (even at the risk of repetition) the importance *both* of the highest sense of individual responsibility *and* of governmental use of the guidance of nutritional knowledge.

EXPERIMENTAL SCIENCE AND TRADITIONAL ECONOMICS

Both in personal and in public nutrition policy, there is often implicit, in the practical questions that arise, a far-reaching general problem. How far do the food habits which are traditional with us represent the outcome of a racial experience which can safely guide our use of food resources and to what extent should our food habits now be brought under the guidance of the present-day science of nutrition?

At the risk of attempting the hardest step first, let us begin with the tradition which mortgages the outstanding share of the food money in most American households.

Liberal meat eating long ago gained and still holds a prominent place in our food traditions; it symbolized success in the chase and in pastoral pursuits, prosperity, liberality in hospitality, and is the central feature in most of the folklore and history of feasting, the traditional "main dish" of the dinner.

In historic times, the people of many regions have lived largely upon grains, tubers, and roots; and today, because of economic limitations, too many people are living too largely upon these foods.

It seems to be accepted as a fact of observation that in the present general stage of human history (perhaps for 20 to 40 generations including our own) people who of necessity are living very largely on starchy food show a tendency to eat more protein when their economic conditions permit.

This has been called *the protein shift,* and some economists regard it as a well-established phenomenon which has occurred in many countries either simultaneously or successively as their economic conditions have permitted; it can often be seen in progress as different economic strata of the population of a given country succeed in improving their standards of living. Here tradition was visibly taken over into conventional economics and dignified with a technical name. Unfortunately, however, this traditional-economic rationalization was misleading. It fostered the view that, given sufficient food calories, "protein" was the one desideratum. This mistaken view retarded and still retards an understanding grasp of the newer knowledge, won through experimental science, which shows that the usual shift of food habit with increased purchasing power enriches the dietary in mineral elements and vitamin values as well as in protein contents.

Moreover, the present-day science of nutrition throws the emphasis of the shift less toward meat as the typical high-protein food and more toward a combination of milk and fruit with its better balance of protein, mineral, and vitamin values.

Thus in his official Foreword to the United States Department of Agriculture's Yearbook for 1939 Secretary (now Vice President) Henry A. Wallace wrote:

Probably 99 per cent of the children of the United States have a heredity good enough to enable them to become productive workers and excellent citizens provided they are given the right kind of food, proper training, and ordinary opportunities. Fundamental to adequate training and decent opportunity is food. Fifty per cent of the people of the United States do not get enough in the way of dairy products, fruits, and vegetables to enable them to enjoy full vigor and health.

The increased prominence which Wallace gives to dairy products, fruits, and vegetables in the American food supply as a matter of nutrition policy is in full accord with the teachings of such scientific leaders as McCollum and the late Dr. Mary S. Rose.

In a food budget guided by the newer scientific knowledge of nutrition, the prominence of fruits, vegetables, and milk would certainly be increased and that of meat probably somewhat decreased as compared with their respective traditional places in the American food supply.

When the influence of the newer scientific knowledge tends to assign to the product of any given food industry a less prominent place in the food budget than that which tradition would give it, the traditional-economic viewpoint may set itself against that of the modern experimental science of nutrition, claiming perhaps that the science of economics is more realistic than the science of the laboratory.

From the traditional-economic viewpoint the argument would be to the effect that the "real" place of any commodity in the food budget is that which it holds and has held rather than any lesser place which nutritional research might indicate. Traditional arguments and emotions including those of social custom and prestige and the often misleading concepts as to "standard of living" may all play a part in maintaining the consumer demand for, and so the price of, a food at a higher level than would correspond to its nutritive value. Yet laboratory science can now determine nutritive values quite objectively; it has shown that these are the real values so far as the well-being of the consumer is concerned, and possess potentialities for constructive benefit such as were not known or conceived at the time of the formation of the food traditions hitherto held. As has been pointed out by no less eminent and judicious an authority than Hopkins, traditions in the food field tend to accumulate prejudices quite as often as truths, while the present-day science of nutrition, built on the experimental and exact sciences of physiology, chemistry, and physics, is objective and its findings have been shown to be valid and constitute a trustworthy guide for the advancement of human welfare.

CONSUMER DEMAND AND AGRICULTURAL ADJUSTMENT

In the United States, the Federal and state governments have made many contributions to nutritional research and education, but as yet have taken only tentative steps in the way of direct economic action to increase the proportion of the most "protective" foods in the nation's food supply.

Governmental agencies teach the benefits of such a higher proportion of protective food, but the actual bringing about of this improvement seems to rest mainly upon the basis of consumer demand. Educational guidance of consumers to a larger use of protective foods thus becomes an important part of nutrition policy in the United States.

The phrase "protective foods" as here used means, as explained in a previous chapter, fruits, vegetables, and milk (including cheese, cream, and ice cream), with or without eggs. It leaves wide latitude for individual preferences and for due consideration of economic conditions in the choice among fruits and vegetables; and also for choice among the different forms of milk (fresh, canned, and dried) and such of its products as sufficiently possess its nutritional characteristics— fermented milk, cheese, cream, and ice cream.

Intelligent and scientifically informed consumer attitude is an extremely important factor in our progress toward improved nutritional well-being for all. This progress may be either helped or hindered by the food choices of those whose purchasing power is higher than that of the average consumer.

Within living memory citrus fruits were a luxury. People of sufficient income to afford it took up the habitual use of these fruits, making for them a market which encouraged larger plantings with resultant economies of production and distribution, lower retail prices, and increasingly wider consumption. As a result, the oranges and grapefruit which a generation ago were available only to people in the upper income brackets are now used habitually through long seasons by people of all levels of income. Science shows that this is a good investment for them all at the prices which the present development of the industry makes possible.

Somewhat similarly, the development of the cold-storage industry a generation or so ago was made economically practicable by the willingness of well-to-do Americans to pay the high prices which then prevailed for winter eggs. As the extension of commercial cold storage acted to eliminate the season of scarcity, eggs became a practicable year-round staple for a much larger proportion of the people.

Still again, it was the slowly developing appreciation of milk as a food, first on the part of people who could afford relatively high prices in winter and for superior sanitation on dairy farms, which resulted in such development of the market-milk industry as has made a high quality of milk a year-round staple for the poor as well as the rich.

In some respects the meat industry has seen a similar evolution, but in other respects it presents a fundamentally different economic situation. Under pioneer conditions the large areas of grazing land not yet brought under the plow, and sometimes also abundance of game, make meat a relatively cheap food. People coming from the old world to a new world of open frontiers could expect not only better chances to prosper through enterprise but also that even from their first arrival in the frontier country they could eat meat oftener than in the old countries from which they came. But with the taking into cultivation of what had been free-range frontier pasture lands, a point was reached beyond which meat production could be increased only by feeding to meat animals, in addition to their pasturage, a part of the field crops of the cultivated land. This makes the meat more expensive, and correspondingly there has been built up a tradition that grain-fed beef is worth more per pound than that of an animal marketed directly from range pasture.

The traditional associations of the level of meat consumption with standard of living and of grain-fed properties with quality in meat together result in a consumer demand for relatively large amounts of meats which are inherently costly to produce. The price is not felt to be unduly high because the market expert gives such meat a high conventional "grade," and because it is expensive to produce. Unfortunately its production costs more, not only in terms of cash but also in terms of the country's resources. If consumer demand for grain-fed meat were moderated, if meat were made more largely on range pasture, and the farmer fed his field crops more largely for milk and egg production, the same food-production resources would result in a better nourished people.

As explained in a previous chapter, we interpret the scientific evidence now available as indicating that for the best results as much as half of the needed food calories should be taken in the form of fruits, vegetables, and milk in some form or forms. If all consumers understood this advance in scientific knowledge and acted upon it there would be increased consumer demand for these protective foods, and undoubtedly agricultural production would promptly adjust itself to this shift of demand.

The United States can readily produce on essentially the same lands now devoted to food production, the additional fruits, vegetables, and milk (perhaps also eggs) which are needed for the better nourishment of its people. If a small percentage more of land were wanted it could easily be diverted from the raising of cotton for which the full acreage of the past is no longer needed.

The United States Department of Agriculture Yearbook for 1939 collects many of the materials for the construction of a national nutrition policy. Its pages and other writings of the experts of that department leave no doubt that our farms could readily double their present production of practically any fruit or vegetable for which there is sufficient consumer demand. Fruit occupies only from 1 to 2 percent, and vegetables only about 4 percent of the crop acreage in this country, so that the total production of both could be doubled or trebled without affecting the returns from other crops to any serious degree. Land and labor formerly devoted too exclusively to cotton, and for which diversification is now universally recognized as needed in any case, may well be devoted, in part or in rotation, to the growing of small fruits, melons, and fresh vegetables ("truck crops"). Here there is an enormous potential resource for the meeting of a growing market demand for fruits and vegetables.

Growing appreciation of fruits, vegetables, and milk is shown by comparative dietary studies at a fifteen-year interval in New York City, by the Federal marketing statistics, and by the reviews of food-consumption data in the United States Department of Agriculture Yearbook for 1939.

Nor is it only in this country that the production pattern is gradually shifting to meet the consumer demand for fruits, vegetables, and milk. Speaking for conditions generally in the Western World, the 1937 Nutrition Report of the League of Nations states that the best evidence that agriculture can and will adapt its production to the nutritionally guided consumer demand is that it is doing so.

It is also reported that when the United States Department of Agriculture was conducting milk-for-health campaigns it appeared that these had at least as immediate an effect upon production as upon consumption, so that the milk business grew but yielded no increase of profit to the producers. Even so, however, the shift toward increased milk production is a step in the normal evolution of American agriculture which puts the farmers who take it in a scientifically sounder position for the long run, because dairy cattle yield more of nutritive value to the human food supply (in proportion to what they themselves have eaten) than do any farm animals that yield human food only by slaughter.

A fact which needs increased recognition in nutrition policy is that the dairy cow, even after she has passed her peak of milk production, is a more efficient converter of field crops into human food than is any animal which contributes only meat to our food supply.

Perhaps this will be the more readily accorded if it is also emphasized that recognition of this fact does *not* foreshadow a meat-less food supply. In all the major regions of the United States there are large areas of grazing land which produce grass-fed meat and cannot advantageously be turned to any other use. Moreover, every farm which produces milk or eggs produces meat also. Undoubtedly meat will continue to occupy a prominent place in the American food supply: the only question is, How prominent? And this question nutrition policy should raise with regard to every major item of the food budget.

With growing knowledge of nutritive values, we learn to see more clearly what constitutes a wise allocation of the country's food-producing resources and of the money which the consumer spends for food. We have seen, too, how the evidence now makes it plain that in some lines of food consumption the purchasing power of the well-to-do may serve to make possible the up-building of an industry which as it develops improves the dietaries of the poorer people as well; while in other lines, a consumer demand for high-priced food works permanently against the prospect of good nutrition for all.

Moreover, in this respect what is true as between the richer and poorer consumers in any given country is also true in some degree as between the "have" and "have-not" nations.

Yet the tension between nations can be greatly relieved by the application of the same scientific knowledge of nutrition which reveals the inevitability of such relationships as we have just been considering. For now that food traditions can be evaluated by objective scientific evidence, and the qualitative and quantitative nutritional values of the foods which flourish in different parts of the world can be assessed, international exchanges of things involved in human well-being can go on with greater confidence and mutual goodwill.

INTERNATIONAL RELATIONS AND THE VIEW AHEAD

Wheat is so abundant, and the prospects for its continued abundance are so good, that the problem of bread (viewed in the large) seems chiefly a combination of distribution problems. Conspicuous among these are the problem of available ships to transport the wheat or flour, and monetary means of supporting an adequate flow of breadstuffs from the original producer of the wheat to the ultimate consumer of the bread.

The United States, Canada, Australia, and New Zealand have large reserves of wheat and can easily increase their production to provide for the feeding of the countries that need to import. The same will doubtless be true of Russia after her invaders are expelled and the damages inflicted by the invasion are repaired. The Argentine will doubtless also have surplus wheat.

The degree of peacetime exchange of foods and services now contemplated as mutually and permanently advantageous by the United Nations can, should, and presumably will mean that normally no population groups shall lack bread.

What beyond the grain crops will each nation need to make adequate its food supply? The answer depends somewhat upon our angle of approach. Beginning at home, the most useful start for those who read this might well be the principle that, "The dietary should be built around bread and milk." But in some countries there is such pressure of population upon the food supply that the people must live mainly on the primary products of the land. If then we consider first the foods which can be obtained directly from the land, it is the green-leaf vegetables and the yellow-fleshed fruits and vegetables which best supplement the grain crops by supplying the calcium and vitamin C content and the vitamin A value which our nutrition needs in larger measure than the grain crops supply them. In the densely populated parts of China most of the people are obliged to depend chiefly on leaf-foods to balance the rice, wheat, or millet which is their main source of calories and protein; and so they must eat greens in quantities which would seem burdensomely bulky to us.

So favored in food-production resources are we of the United States that normally we can feed to cattle, as transformers of green leaves, hay, and grain into milk and meat for us, the greater part of our total grain crop and a vastly larger proportion of the green stuffs that grow on our farms. As already noted there are parts of this country where the farmer has definitely to decide whether to turn his field crops more largely into milk or more largely into meat. Yet there are also large areas of pasture lands not so situated as to be suitable for dairy farming and which do ensure a liberal meat supply for an indefinite future. An increase of 50 percent in our milk supply would probably have the effect of decreasing our meat supply by only 5 to 10 percent.

For a long time at least we shall be able to share both meat and dairy products with other nations while at the same time improving nutritional conditions at home. In the Argentine there are great areas of grazing land which (even though part of it ultimately be gradually brought under the plow) will in all probability produce beef for export for as long a future as we can foresee.

The typical American dietary as observed in all the major regions of the United States and at all economic levels is most improved by increasing its proportion of fruits, vegetables, and milk. In contrast, the foods of which we produce largest surpluses are wheat and meat; and others which we also may well share more liberally with other nations, by letting more of our accustomed imports go to them instead, are sugar and fats.

Thus a scientifically sound nutrition policy can consistently contemplate the improvement of the nutritional well-being of the people of the United States at the same time that we share our food supplies more generously with other people.

At the time that this is written we are explicitly committed to the policy of making common cause with Great Britain in the matter of food supply. For this reason and also because the normal prewar patterns of food consumption were much alike in the two countries, the British war experience has intimate bearing upon the food problems that face us, both at present and for the near future.

Both countries are well supplied with wheat and potatoes; but Great Britain has been, and because of shipping conditions still is, distinctly short of fats and meat compared with her levels of prewar consumption. The British also greatly welcome our shipments of dried eggs, dried skim milk, cheese, and canned and dried fruits. These foods are also much wanted in Russia and China and the policy of common cause which is explicit between us and the British must obviously be regarded as more or less implicit in our relations with Allies to whom we owe so much as we do to the Russians and the Chinese—though of Russian and Chinese food conditions we have less accurate knowledge and their prewar dietary patterns were not so closely like our own.

For years to come it will be important to keep in mind that people who are in process of rehabilitation from the weakening effects of privation and anxiety are not yet ready for drastic changes from the food habits to which they grew up. In sharing our abundant grain crops with the peoples of other countries who have lacked bread, or that equivalent which is *their* time-honored staff of life, we should take account of the food customs of those to whom we send. Our rice crop, for instance, might well be divided between certain limited sections of the United States and certain much larger areas in China. Other parts of China are sufficiently accustomed to the use of wheat to make it their main food. In Russia different regions specialize to some extent in different grains as best suited to their conditions of climate and soil; but wheat is welcome everywhere and advantageously used so far as available.

Only in limited regions of China and Russia is maize a sufficiently familiar food to make it wise for us to send large quantities to them. But our farmers are well accustomed to turn their surplus corn crops into pork, and all the pork that we can ship will be welcomed and well used by many millions of people in all parts of the world.

Eggs also are a food which (even when dried before shipping) are a welcome and beneficial enrichment of food supplies almost everywhere. Milk is not so universally familiar as a major food for people of all ages. While we believe that ultimately all people will profit by a per capita milk consumption as high as our own, it may be wise to begin by sending our available surplus of milk (1) to those who distinctly desire it, and (2) to the particular places where nutritionally minded people can promote its use, first, in child feeding. Dr. J. C. Thomsen, a professor of chemistry in Nanking University who has especially studied the food and nutrition problems of China, reports in a personal communication that even in parts of China where milk was hardly known as a commodity, the people have readily learned the use of imported dried milk in the feeding of their children. Both New Zealand and the dairy regions of our Pacific states are large potential sources of dried milk for the Orient as well as for all of their present consumers.

In so far as frank and friendly competition for markets leads food producers to increased efficiency in the production of nutritive values, mankind will benefit. But as each nation's nutrition policy should recognize that some foods have values which are above price because they help to make possible the improvement of the quality of human life, so also we may hope that food policies as between nations shall look not only to money profits but still more to such sharing of the fruits of the earth as to make for goodwill and a just and lasting peace.

Nutrition policy may be a large factor in helping the world to learn to think in terms of good lives for all its people.

CHAPTER XV

SCIENTIFIC CRITIQUE OF THE "OFFER" OF HIGHER HEALTH AND LONGER LIFE

FOR well over two decades, McCollum, on the basis of his outstanding work in biochemistry, has taught that there may be important differences between the merely adequate and the optimal in nutrition; and J. F. Williams, doctor of medicine and authority in physical education, has explained that health may and should mean not merely freedom from disease but rather a positive quality of life, which can be built to higher levels.

Although after some years of such teaching, Dr. Williams remarked that this view had not found wide acceptance, yet today we see many signs of a growing appreciation of the fact that there are different degrees or levels of positive health; and undoubtedly nutritional research has had the major share in the advancement of this ideal of optimal as distinguished from merely passable health. For the objective and quantitative nature of the findings of much recent experimentation carry an impersonal convincingness which advances *the principle of the nutritional improvability of the normal,* out of the realm of opinion into that of established scientific fact.

Nutrition thus becomes at least coördinate with sanitation in the service of human health; but whereas sanitation is brought into service chiefly through public-health practice, nutrition is more largely dependent upon conscious choices by individuals and families. In a few directions and up to the level of controlling obvious deficiency disease, nutritional benefits may be administered wholesale after the manner of public sanitation. Examples of this are the bread-enrichment program described in previous chapters and the extent to which pellagra has been reduced by the large-scale distribution of yeast in the south (DeKleine, 1942). But the greater number and especially the higher levels of health benefits available through nutrition can not be conferred automatically upon a merely passive public. In the main, scientific discovery *offers* these nutritional benefits to those who will actively use the guidance of the newer knowledge in their daily food habits. The benefit depends upon the individual *will* to use the knowledge.

That being the case, the reader may wish to be shown in a somewhat more critical manner a selection of samples of the original evidence upon which science bases its offer of higher health and longer life to those who elect to use the guidance of the newer knowledge of nutrition. Such is the purpose of the present chapter.

If it should seem that we here deal largely with data derived from experiments with animals, the reader is asked to keep in mind that so far as it goes, the evidence of direct human experience is entirely in harmony with the findings here reviewed. It need not be argued that scientific critique can be more rigorous when applied to data obtained in larger numbers and under more comprehensive laboratory control than is possible with human subjects.

Experiments in Terms of Natural Foods

It is a commonplace of research that experiments should be so planned as to introduce only one variable at a time. In nutritional research these experimental variables are of two kinds: (1) individual chemical factors—elements or compounds as the case may be; and (2) the actual articles of food which nature and agriculture produce and which people obtain and consume.

The protective foods, being more expensive to produce and distribute, must cost the consumer more per thousand calories than, for instance the staple cereals and breadstuffs. Hence it is important to know how much protective food is needed to balance the dietary or food supply. The answer to this question was sought experimentally in the chemical laboratories at Columbia, making use of animal-feeding experiments carried out with as rigorous a regard for exactness as in the most careful of quantitative analysis *in vitro* (for, as a matter of fact, it was as an extension of quantitative analysis of foods that the Columbia chemical laboratory took up such feeding experiments). We found that within the range of individual (physiological) variation of the zone of normal nutritional response there was a conclusively measurable trend to higher average records on the part of the individuals whose dietaries contained higher proportions of protective food.

Thus, as has been briefly mentioned in Chapter XII, Diet A (a mixture of one sixth dried whole milk with five sixths ground whole wheat plus table salt and distilled water) was adequate to enable our rat families to live successful lives generation after generation; but better average records were made by strictly parallel families of the same heredity when fed Diet B which differed only in containing a higher proportion (one third instead of one sixth) of milk powder in the wheat-and-milk mixture.

It is hardly necessary to explain that the significance for human nutrition of such a comparison as this lies in the fact that these experimental diets *differed from each other in the same way* that human dietaries often do. With the *difference* the same, the study of the effects of this difference is valid for our purpose. The fact that the experimental dietaries contained fewer articles of food than human dietaries usually do makes no difference to the validity of the findings and is an advantage to clarity of interpretation. The question at issue is the nutritional improvability of the normal. Diet A supports normal life throughout the life cycle and through successive generations; but an increase in the proportion of milk in this diet gave statistically conclusive evidence of improved well-being at every stage in the life history. The experiments were begun by assigning the two diets to equal numbers of carefully matched laboratory-bred rats 28 days of age (this being conventionally taken as "end of infancy" in the rat).

In his omnivorous food habits and in his nutritional processes the rat resembles the human being quite closely. Only two noteworthy differences are known in the nutritional chemistry of the human and rat species, namely, that human beings are much more responsive than are rats to the vitamin C and to the niacin contents of their diets. Hence all the favorable effects of superior nutrition upon health and longevity demonstrable in the rat occur likewise in the human body, while the latter benefits still further when the dietary is also improved in these factors to which the rat is indifferent. So the nutritional improvement of the norm which the laboratory evidence reveals is *safely and considerably within* the actual scientific probabilities of the benefits of analogous dietary improvement in human experience.

Furthermore, in order that the reader may not be left in any doubt, it may be emphasized at this point that throughout this book all findings cited or views offered are in accordance with the evidence of direct human experience so far as this latter evidence goes. But often the evidence from experiments on animals used as "deputies for humans" is much fuller and more completely controlled.

A paragraph follows on the findings with reference to each of the chief objective criteria used in this comparison of Diets A and B.

Rate of growth.—The matched lots of 28-day-old rats assigned in parallel to Diets A and B were first compared with reference to their individual gains in body weight during the 5th to 8th weeks, inclusive, of their age. Both males and females averaged higher gains on Diet B than on Diet A, the differences being more than 25 times their probable errors. This degree of statistical conclusiveness leaves no room for doubt. Also, this finding (published in 1924) met no special inclination to skepticism because it was considered common knowledge that children, even starting from within the normal range, grow better when better fed; and a little later this was found in well-controlled work with children, by Mann and in the Milk-for-Schools Scheme as has been noted in Chapter XII.

Efficiency of growth.—The gains in weight of the young rats receiving Diets A and B respectively, as explained in the previous paragraph, were divided by the energy values of the food they had consumed, and averaged to show the efficiency of growth in terms of body weight gained per thousand Calories of food consumed during the 5th and 8th weeks inclusive of the rats' lives. This corresponds approximately with the few years of most rapid growth in children. A preliminary series of experiments indicated that there was no need of distinction between the sexes in applying this particular criterion of the relative nutritional efficiency of Diets A and B, so the main series on which this point was studied consisted of 100 experimental lots, each a family group of 3 females and 2 males, on each of the Diets A and B. The efficiency of growth on Diet B was found to be distinctly greater, the difference being more than 16 times its probable error. Actually the degree of statistical conclusiveness is higher than the last sentence implies, for in its computation of probable error, the family lot of 5 animals is counted as a unit "case" so that each average actually represents 5 times as many individuals and its actual probable error is correspondingly smaller.

Average size.—At any given age the animals of either sex on Diet B average larger than those of the same sex on Diet A. All the averages were, however, well within the normal range for animals of our laboratory colony of the respective sex and age. There was no appreciable difference in the proportion of body fat nor in the apparent relation of skeleton to musculature, but quantitative analysis of parallel animals showed that not only the actual amount but also the percentage of body calcium averaged higher for a given age and sex among the animals on Diet B. As is explained elsewhere, this influence of the diet in expediting the normal processes of calcification in the growing body has been definitely correlated with better average life histories. This is significant, whereas the mere fact of higher body weight at a given age, while undoubtedly real (statistically significant), might in itself have no bearing upon the nutritional well-being. It is hardly necessary to point out that in the scientific view of today, to be better nourished does not mean to be bigger or fatter, but rather that the life processes go on in a nutritionally more favorable bodily environment.

Time required to reach maturity.—In order to have an entirely objective measure of the time required to reach maturity, and of whether, if two diets support different rates of growth, they are supporting different rates of development also, the young rats of the two sexes were allowed to grow up together, three females and two males in a cage, and the age at which each female gave birth to her first young was made a permanent part of the experimental record. In the same comparison of Diets A and B to which the preceding paragraphs have referred, it was found that Diet B not only supported somewhat more rapid and more efficient growth, but at the same time expedited bodily development, young being born on the average earlier to the females on Diet B than to their twin sisters or double cousins on Diet A. The difference between the averages was over 20 times its probable error, giving a very much higher degree of statistical conclusiveness than even the most rigorous standards of interpretation require to justify the term "undoubted." As a matter of fact, this finding, published in 1924, has remained quite unquestioned and has been frequently confirmed in later experiments. It will be noted that the same dietary difference which expedited growth and the normal calcification of the body resulted in earlier maturity in a parallel way. This fact is to be taken as one of many evidences that the reproduction record of the experimental animals is a valuable criterion of their nutritional well-being. If these reproduction records should seem to the reader to occupy a relatively prominent place in the experimental investigations of nutritive values of dietaries, it is to be remembered that they are so employed because they afford data of entirely objective kind and not because we are dealing with any specific relations of the diet to the reproductive process.

The period of full adult capacity.—This is symbolized in experimental animals by the duration of reproductive life. When one keeps in mind the fact just mentioned, the duration of reproductive life becomes another important criterion of the relation of nutrition to general well-being. Relatively early in the development of the new science of nutrition, McCollum, Simmonds, and Parsons pointed out that diet may influence the onset of senility. McCollum, in the successive editions of his book *The Newer Knowledge of Nutrition,* and many other recent research workers in scattered articles written from different points of view, have cited what, in total, amounts to a strong body of evidence that the food habits of individual people and the differing food supplies of different nations do influence markedly the number of years between the attainment of maturity and the onset of senility in human life. While, as just said, it is still somewhat scattered —and this is true both of times and places of publication and of the terms in which the observations directly upon human beings have been recorded [1] —yet it seems fairly clear to those who find time to study the evidence that large numbers of our species have already demonstrated in their own persons that the kinds and relative proportions of the foods we consume do influence our human life histories, and that this is as true for those above as for those below the general average status

[1] Just as this is being written, there comes to hand an article by Dr. V. Korenchevsky, published in the *Journal of the American Medical Association,* 1942, 119: 624–630, under the title, "The War and the Problem of Aging" which, on reading, one finds to relate very largely to phenomena commonly spoken of as those of senility but in which nutritional status is now known to play a large part. We have also seen in Chapter X that the recent researches of Kruse have shown and are increasingly showing how large a part long-continued suboptimal nutrition plays in bringing on the bodily changes commonly called senile.

of bodily well-being and the general average duration of the prime of life. While the scientific viewpoint does not doubt this fact, it also does not see any prospect of obtaining accurately controlled observations upon human beings in such numbers and over such lengths of time as would constitute evidence of comparable conclusiveness and objectivity with that obtainable from long-established, laboratory-controlled colonies of experimental animals of properly chosen species. Hence McCollum and his coworkers at the Johns Hopkins School of Public Health, in addition to their studies of the literature of human experience, have made use of their colony of laboratory rats for experimental comparisons of different diets with respect to the influence of the nature of the diet upon the time of onset of senility. They have especially remarked the suitability of the rat because (in addition to the advantages already mentioned) of the nature of the ways in which the rat reveals senility by outward appearance and behavior. The experience of the nutrition-research workers in the Columbia Department of Chemistry has confirmed and extended the findings just mentioned. In the Columbia work, it has also been deemed desirable to make use of some criterion which would be regularly recorded in numerical terms and which should be entirely objective (entirely independent of the judgments of persons). For while we at Columbia are in full agreement with the Johns Hopkins investigators as to the signs of senility in our experimental animals and the interpretation of the significance of these signs, yet so far-reaching is the importance of the human implications of the nutritional improvability of our normal lease of fully effective life, that complete objectivity here has a special value in making the findings convincing to a wider scientific public. Thus objectively measured and then interpreted by equally impersonal statistical analysis, the data of the above described comparison of Diets A and B show that the latter with its higher proportion of milk resulted in an average prolongation of the period of full adult capacity to an extent undoubtedly of high significance, the difference being 11.6 times its probable error. Thus the impersonal convincingness of the scientific evidence by which this finding is supported is of a decidedly higher order than that unquestioningly accepted as establishing most others of the fundamental facts of physiology. If one has a more tentative feeling about the principle of the nutritional improvability of the norm than about other principles of physiology it is not because the scientific status of the evidence for nutritional improvability is less conclusive; but only because, being more recent, it may not yet have been assimilated into one's everyday working state-of-mind.

Adult vitality as indicated by success in the launching of, successive generations.—Here again it is to be kept in mind that in animal experimentation we make use of reproduction records because they are the most objectively measurable of the criteria of adult vitality, and not because the differences in diet which we are studying have any specific relation to reproductive function. Nor should economic or sociological predilections be allowed to prejudice us subconsciously against the validity of this mode of investigation. A family which does not wish to have more or larger offspring may and presumably does have other objectives toward which to apply that same superiority of nutritional well-being and internal environment which in the experimental animal finds quantitative expression in the number and vigor of the offspring which it contributes to the succeeding generation of its kind. Men and women who are not interested in adding to the numbers or to the bodily size and vigor of the next generation may nevertheless be very much interested in the higher degree of buoyant health which those criteria symbolize in scientific experimentation. Both Diets A and B support successful reproduction and rearing of young, generation after generation. The families on Diet B, however, reared more young, and to a better size and development at the conventional uniform "weaning" age. The numerical differences on these two points were, respectively, 13.7 times and 10.5 times their probable errors, or thousands of times more conclusive statistically than the accepted criteria of scientific interpretation deem sufficient to justify speaking of the reality and significance of the difference as undoubted.

Length of life.—Only after the general method of research described and illustrated in the preceding paragraphs had thus shown its usefulness and high scientific validity were means forthcoming to continue controlled feeding comparisons throughout the entire natural lives of experimental animals. Ultimately, however, parallel animals of both sexes were continued on each of the Diets A and B from the end of infancy until natural death up to totals of 124 to 196 individuals of each sex on each diet. In essentially equal proportions of cases on the two diets, "second (or later) generation" animals were used. That is, the animals thus assigned to the test (as contrasted with the "first generation" cases where litter mates were assigned to the two diets equally) would be, for example, double first cousins having exactly the same hereditary background but with a nutritional background of a previous generation on either Diet A or Diet B as the case might be.

The purpose in this plan was to give an adequate number of fresh starts with direct litter-mate controls and also sufficient representation to cases in which the diets were compared for their effects in a second generation because effects of dietary differences are sometimes clearer in the second generation than in the first. In contrast to many studies of longevity in which exceptional cases are sought out and enquired into, this was a study of the influence of a dietary difference (that between the above-described Diets A and B) upon the average lengths of life of the general population. More strictly speaking, it was the average lengths of life of presumably adequate and representative samples of the entire population except those dying in infancy. As death rates both of our present-day population and of our laboratory animal colony are relatively low for a considerable period after the completion of infancy, the length-of-life as studied in this comparison of diets corresponds fairly nearly to the "adult life-expectation" of human vital statistics. This average length of life was almost exactly 10 percent longer, both for males and for females, on Diet B than on Diet A, and the differences were, respectively, 5.5 and 5.9 times their probable errors. This means statistical probabilities of the order of 10,000:1; or that the finding has 100 times higher conclusiveness than we would need to justify calling the conclusion undoubted.

As a matter of fact the finding is considerably more firmly established than this statistical examination of the averages by itself implies. For the data may also be evaluated in another way, in which with the data set up in terms of percentages of cases reaching certain well-distinguished age goals, we have nine different comparisons of the diets and in each of the nine comparisons, Diet B shows a measurably better result than Diet A.

Moreover, a preliminary comparison, made when about half the cases had been completed, showed the same 10 percent difference in favor of Diet B and essentially the same coefficient of variation. The final numbers, therefore, presumably constituted a more than adequate sample to establish the true difference, and doubtless if more experiments of the same kind had been made, the probable errors of the means and of the differences would have become still smaller and the statistical conclusiveness higher; but, with a degree of conclusiveness 100-fold better than "undoubted," it seemed that the right use of the remaining time and opportunity would be to get added light from a different but coördinated type of experiment.

The increase in length of life through nutritional improvement of a diet already adequate and a nutritional status already normal, as here described, has been accepted as undoubted since its publication in 1930, although it has been more discussed because more unexpected than the other findings of this series. Its unexpectedness is largely due to the fact that it importantly extends the answer to a question which our half-knowledge had supposed to have been answered already. It had been supposed that "heredity determines longevity" and that "while of course you can do things which will shorten your life, the only way you can lengthen it is by the selection of a longer-lived ancestry." The first of these oft-repeated dogmas represents roughly about one half of the truth. According to our present knowledge (in the light of the feeding researches here summarized and others still more recent) it is clear that *heredity and nutrition are both major factors* in determining the length of life. The influence of nutrition is now established with much the greater completeness of scientific control and statistical convincingness of the two, yet we do not doubt that heredity is also influential. The old ways of speaking as if heredity were the only major influence should be entirely discontinued because they are so seriously misleading; but the correction of misleading half-truths is not the same as the displacement of one theory by another and contradictory theory.

Whether done unconsciously or in a deliberate attempt to "dramatize" or to express in a "sprightly style," it is an unscientific habit to speak of every fundamental advance in scientific thought as if it were in conflict with some previous theory. Our recently acquired knowledge that nutrition influences the length of life (and can influence it for better as well as for worse) *does not constitute a rival theory* but simply makes an important addition to our previously acquired knowledge regarding the influence of heredity. These two kinds of knowledge are acquired by such different methods of research that as yet we have no conclusive ways of measuring the relative potencies of the two influences, and indeed there is strong (inferential) scientific probability that the one influence is relatively stronger in some cases and the other influence in other cases. Also there is scientific reason to suppose that heredity is relatively more potent in those cases of *extreme* longevity which seem to "run in the family" much as extreme height may run in the same or other families. Yet it may also be true that in a much larger number of families there is no such extreme hereditary trait, and that consciously cultivated, nutritionally guided, good habits as to what foods one eats and in what proportions may very probably be the greatest constructive influence contributing to higher health and longer life. It is also well worthy of emphasis that whatever one's original chromosomal endowment and whether his hereditary background is on-the-whole an asset or a liability, he still, as a reasonably normal member of a mammalian species, influences through his food and nutrition the development of his innate potentialities, either for better or for worse.

Let us now look briefly but critically at the evidence of nutritional research planned in terms of the individual chemical factors or specific nutrients.

Experiments with Different Levels of Protein Intake

Slonaker studied at Stanford University the comparative effects of diets of the same general type, but in which the percentage of protein was set at 5 different levels: 10, 14, 18, 22, and 26 percent, respectively, of the air-dry food mixture. These differences in protein content were brought about by varying the percentage of lean meat in the food mixture; so it was not quite exclusively a protein difference, but we may follow the original investigator in treating the findings in terms of protein. The five different levels of protein intake did not always arrange themselves in the same order with reference to the different physiological data which Slonaker considered as possibly throwing light upon bodily well-being. Gain in weight increased with the percentage of protein in diet; but spontaneous activity showed no definite trend, though the group having highest percentage of protein in the food was least active. Males lived longest on the diet with 14 percent protein; females, with 10 percent; both showed shortening of life on diets containing 22 or 26 percent protein. The reproductive span showed no regular relationship to the protein content of the diet. Here the only noteworthy regularities would seem to be that with increasing protein content of diet, growth was faster and life was shorter. The length of life was, however, not conspicuously different. That life should be somewhat shorter after the forcing of early growth by high protein diet is in line with the findings of Maynard and McCay at Cornell, whose animals, however, showed larger differences in longevity than did Slonaker's.

In experiments at Columbia also, the addition either of meat or of purified casein to an adequate diet has resulted in more rapid growth to a somewhat larger adult size, but no clear evidence of higher adult vitality, and certainly no increase in the length of life. This general picture of the effect of adding protein to an adequate diet was no different when the percentage of calcium in both the diets was higher. Also it was found that the more rapid growth in body weight which results from higher protein was not accompanied by any increase in the rate of development as indicated by the percentage of calcium in the body at a given age.

Thus the general results of the Stanford, the Cornell, and the Columbia experiments consistently indicate that high protein intake increases the rate of growth but does not conduce to higher health or longer life. Hence we conclude that in the above described comparison of Diets A and B, protein may have been an appreciable factor in the more rapid growth and increased body weight on Diet B, but could have played but a minor part, if any, in the improvement of nutritional well-being which resulted in a life history of higher health and increased longevity. It is, however, quite conceivable, in view of the relationships to tissue enzymes mentioned near the end of Chapter VIII, that protein enrichment simultaneously with just the right enrichments with phosphate and some of the B vitamins might be made to play a larger part, and this problem is now being studied at Columbia.

Experiments with Different Levels of Calcium Intake

When, without any other change in the above-described Diet A, its calcium content was increased to the same level as in Diet B, the results showed that this element had played a large part in the nutritional improvement of the norm which was so fully demonstrated by the comparative feeding of Diets A and B. This increase of the calcium content from 0.19 to 0.34 percent of the dry-food mixture increased the rate of growth and development, the level of adult vitality, the length of life, and, in still larger degree, the length of the period of full adult capacity (the "period of the prime of life"). In all these relations the enrichment of the dietary in calcium alone improved the life processes in the same way, though not quantitatively to the same degree, as the increase in the proportion of milk in the diet.

Thus the *principle of the nutritional improvability of the normal* and the *practical fact of the attainment of longer life with higher health at every stage of the life cycle* are both greatly strengthened now that they have been fully demonstrated in long-term nutrition experiments of both types: those in which natural foods, and also those in which specific nutrients (individual chemical factors) were the experimental variables.

At only one point was the gain which resulted from this moderate increase in the calcium content of the diet too small to have been considered significant if it had stood alone: the average length of life of the females was only slightly advanced. Before concluding that the females were any less able than the males to make full use of the extra food calcium, we have to consider that they had *more ways* in which to invest it; and conspicuously in the nourishment of the next generation during its periods of gestation and of suckling. For, without the aid of any artificial feeding of their infant offspring, the females on the dietary of higher calcium content had produced and reared more young (on the average in the proportion of three to two). Inasmuch as the young rat at weaning represents a considerable investment of calcium, all obtained through the mother, it is clear that the females getting the moderately increased amount of food calcium may have had to invest this limited addition to their nutritional income in their larger contribution to the next generation of their respective families.

This question we have investigated in two ways: (1) The same limited calcium supplement as in the experiments just described was fed (as compared with parallel controls on the basal Diet A) to *unmated* females, and here it turned out that in the absence of the drain of pregnancy and lactation—that is, of the opportunity to invest the extra calcium in the next generation—the females could and did invest an increased calcium intake in an increased length of life quite as efficiently as did the males. (2) Another series of experimental families were submitted to full-life, successive-generation study with dietary calcium as the sole variable, but with a larger increase of calcium in the higher-calcium dietary. The females receiving this still more liberal amount of food-calcium were thereby enabled *both* to rear the already noted larger number of superior offspring *and* to enjoy substantially increased longevity themselves. Thus the individual mother can *both* give more to her children *and* herself maintain and enjoy higher health throughout a longer life *if* the guidance of the newest chemistry of nutrition is followed wholeheartedly and with open mind.

How, then, does one plan for an optimal amount of calcium in one's food? By optimal we mean (here as elsewhere in this book) quite literally the best, and more explicitly the level of intake which induces the best results in permanent practice. As material in hand toward the formulation of an amply adequate answer, we now have large bodies of evidence both (1) from direct human experience including controlled experimentation upon the body's intake and output of calcium, as well as clinical and post-mortem observations; and (2) from well-controlled full-life and successive-generation studies of large numbers of experimental animals with additional litter-mate parallels analyzed for body calcium at selected ages.

Human babies when born (at term) do not differ greatly in the percentages of calcium which they contain—a fact which has been made easy to remember by the somewhat severe aphorism that "the unborn baby is a perfect parasite upon the mother"—meaning, of course, that the chemistry of the situation is such that the embryo develops approximately according to the normal composition of its kind and age almost regardless of whether or to what extent this involves withdrawal of calcium from the mother's bodily structure.

The chemical explanation, or at least its outstanding feature, is that the calcium concentration in the blood, and correspondingly the normal (not necessarily identical) calcium concentrations of other body fluids of the mother and unborn child, are kept near the physiological "saturation" level by the continuous circulation of blood through the vascular ends of the bones, which ensures rapidly renewed contact of the blood with relatively large surface-areas of the calcium salt of the bone substance.

Thus, except for extreme cases in which bone substance becomes badly depleted, fluctuations of intake of food calcium and borrowings of calcium from the mother's bones make little difference to the calcium-concentration level of the fluid, derived from mother's blood, which bathes the developing tissues of the unborn baby. The objectively established chemical and physiological facts leave little room for difference of opinion up to this point; and there is relatively little variation in the calcium content of reasonably normal babies at birth.

From birth onward, however, there is less uniformity either of fact or opinion. Babies are born calcium poor in the sense that the body of the newborn contains a lesser percentage of calcium than the fully developed body. The event of the baby's birth is probably easier and safer both for mother and child because of the fact that the embryo skeleton (even at full term) is still so incompletely calcified. It takes but little scientific insight to see the "survival value" of this species characteristic of being born with bones soft enough to be flexible, for this betters both the mother's and the baby's chances of surviving the event of parturition; particularly when we remember that our species did not evolve physicians and nurses to diminish the hazard of this event until after our anatomy and physiology had become rather definitely stabilized by the processes of biological evolution. Thus, we have no just quarrel with nature for having so evolved us that our children are born calcium-poor, but once the baby *is* safely born, his calcium-poor condition of body is more of a liability than an asset. It would be futile to speculate upon the purely hypothetical question whether or not we should wish, if we could, to raise the young child's percentage of body calcium at once to that which he presumably will have as a normally developed adult, for there is no prospect of that becoming a possible procedure. The real question is whether to content ourselves with a low-calcium dietary (or one in which calcium is left to chance) and a policy of slow drifting toward normal calcification; or to arrange for children to receive dietaries sufficiently rich in calcium and well balanced in other respects to permit them as promptly as practicable to bring their process of normal developmental calcification abreast of their other processes of bodily development and maturation of tissue composition. In the case of calcium, most of the retention goes, of course, toward chemical maturation of the mineral part of the bone tissues. We know that among children ostensibly healthy—and perhaps even among those who might pass as equally well developed in any examination that it is feasible for a school physician to give—the status as to calcium retention and skeletal development may be very different. Along with this difference of actual calcium status among passably normal children goes a somewhat corresponding difference of view as to its significance for health and life history. Some lean heavily upon the assumption that among people passably healthy whatever is, is right; while others believe that what *now* is is the outcome of a period of change and artificiality in our food supply during which there was only partially adequate guidance of nutritional knowledge. To the extent that the latter view is correct, the fuller nutritional knowledge of today can presumably guide us to higher health. And one of the milestones of such guidance is the newly developed understanding of the far-reaching significance of liberal supplies of calcium-rich foods which this section seeks to summarize.

A child may gain in body weight and show a "plus calcium balance" (that is, retain *some* of the calcium he gets from his food) on an intake as low as 0.5 gram of calcium a day. Higher intakes result in higher retentions up to maxima which at present must be expected to vary widely, because the body's capacity to retain calcium depends to such a large extent upon its previous nutritional history and the state of its calcium stores at the beginning of the experiment.

In order to eliminate this source of difference, and so have a more standardized starting point for experiments on the calcium balances of children, some investigators have sought to fill up any possible shortages in the body's normal calcium stores by means of a relatively long period of extraordinarily high calcium intake immediately before making the experiment upon the intake and output intended to reveal the "true" or "net" retention rate and intake requirement. Of course, such a procedure does give a more uniform body status as a starting point for the balance experiment; but it is a status which does not correspond to anything real in the life histories of the vast majority of children. Thus the method of preliminary saturation of the child's body with calcium before studying its calcium requirement seems from the standpoint of scientific critique to be only plausible and not fundamentally sound. Furthermore the practical outcome of such experimentation and of accepting its results as showing children's calcium requirements would be that this artificially construed requirement would be less than the amount that nearly all children actually need. For it is always to be remembered that growing children do not usually have any calcium-saturation period at any time, and if they did they would very soon again have *outgrown* the condition of "saturation." That is, the amount of body calcium which momentarily constituted saturation would tend to be diluted to a shortage by the body's further growth.

A method more consonant with the realities of the problem is as follows: The normal growing child can store (retain) at least 20 percent of his food-calcium intake up to intakes of at least one gram a day. A retention of 20 percent of the amount contained in the food is the approximate average finding of a large amount of work by the University of Illinois group, whose methods of experimentation and interpretation are such as may confidently be expected to result in minimum figures. Jeans and Stearns at the University of Iowa and the Macy group at Detroit have reported considerably higher percentages of retention in experiments which do not appear to be open to the criticism that retentions higher than 20 percent of the intake are to be attributed to previous shortage. There is now a very large uncontested body of evidence that an intake of one gram of food calcium a day is certainly not excessive from the viewpoint of the child's capacity; and that it is easily included in the child's dietary. If children of all ages are given dietaries of suitable natural foods containing not less than one gram of calcium a day, there will result an average retention and building into body structure of at least 0.2 gram a day—more if the body is in a condition of shortage. Such a rate of calcium gain means better development than has been enjoyed by many apparently healthy children in the past. It is clearly a considerable factor in the nutritional building up of the passably normal into the higher health which brings increased efficiency and satisfaction in life. This finding is strongly confirmed by the animal experimentation outlined below.

In our long-term experiments with laboratory animals the level of calcium intake needed for best results, when the entire life history and the launching of the succeeding generation are considered, is not less than two to three times the minimal-adequate level of Diet A which supports the families in health, generation after generation, but apparently with little margin above the bare need for survival of the family.

Table 9 summarizes the records of rats of the Columbia chemical-laboratory colony on Diet A (Laboratory No. 16) and on three others like it in all respects except that they had the higher percentages of calcium shown at the heads of the respective columns—approximately twice, three times, and four times the amount furnished by Diet A.

In this table, the first and second columns of figures are results of a direct side-by-side comparison of Diets A (16) and 162, while the data in the third and fourth columns of figures are from a similar direct comparison of Diets 168 and 169. The fact that there was not an equally close parallelism, in time and litter-mate or double-first-cousin control, between the first two columns on the one hand and the last two columns on the other, is indicated by the double perpendicular line here drawn between them. It is believed, however, that, with the the explanations which follow, the four columns of findings may validly be considered together here. Certainly they are quite as comparable as many data coming from different laboratories or even from the same laboratory at different times which one must consider together in order to take due account of all available evidence on a topic of wide scope such as that to which this chapter is devoted.

The data summarized in Table 9 indicate that, with respect to promptness of reaching maturity and size of young at end of infancy, the "plateau of optimal performance" had been reached at double the minimal-adequate level of calcium intake; whereas with respect to length of life and of prime-of-life (here represented by the duration of the capacity to reproduce) and the number of young reared per female, cal-

TABLE 9

NUTRITIONAL IMPROVEMENT OF THE NORM BY INCREASED CALCIUM CONTENT OF THE FOOD

(Typical Data from Columbia Experiments)

	On Diet A (16) with 0.19% Ca Mean ± P.E.[a]	On Diet 162 with 0.34% Ca Mean ± P.E.[a]	On Diet 168 with 0.64% Ca Mean ± P.E.[a]	On Diet 169 with 0.8% Ca Mean ± P.E.[a]
Age of females at birth of first young............*days*	132 ± 2.4	111 ± 1.6	111 ± 1.0	110 ± 1.0
Duration of capacity to reproduce ....*days*	213 ± 9	279 ± 10	311 ± 10	360 ± 10
Young reared per female	12.7 ± 0.7	20.4 ± 1.1	24.7 ± 0.9	27.0 ± 1.0
Average weight of young at 28 days of age. ....*gms.*	38.9 ± 0.1	42.4 ± 0.1	40.2 ± 0.1	40.1 ± 0.1
Length of life:				
Females ....*days*	723 ± 12	746 ± 13	777 ± 13	824 ± 11
Males ....*days*	658 ± 12	703 ± 11	734 ± 10	729 ± 11

[a] P.E. denotes probable error of the mean.

cium intakes of three to four times that of minimal adequacy were needed for optimal results. A level of intake four times that of minimal adequacy evidently does not overpass the the optimal plateau; for the slightly lower average weight of young at 28 days in the families on Diets 168 and 169 than on Diet 162 is doubtless more probably due to the above-explained less close parallelism. Certainly the general picture presented by the data is most favorable for the highest level of calcium intake.

A less extended series of experiments was made in which the starting point was Diet B, an all-round better-balanced dietary with higher riboflavin content and vitamin A value than the Diet A (No. 16), which was the starting point of the preceding series. Here it was found that with this dietary, superior in so many ways, as the basis, optimal results were more nearly attained with a calcium intake twice that of minimal adequacy and that the plateau of optimal performance was more definitely reached with a calcium intake around three times the minimal adequate level than when the diet was less good in other respects. Here also the highest level of calcium intake tested certainly did not overpass the optimal plateau; rather the higher level is to be regarded as affording the better margin of insurance. That more insurance is needed when the diet is less well balanced is consistent with the findings of the late Dr. S. J. Meltzer that calcium, in addition to performing its own specific functions, serves also as a sort of general regulator of conditions of imbalance in the body.

The series based on Diet A and the series based on Diet B gave entirely consistent pictures of the influence of the calcium content of the food upon that of the body, as may be seen from Tables 10 and 11 (from data given in detail on pages 384, 385 of Vol. 126 and pages 630, 631 of Vol. 137 of the *Journal of Biological Chemistry*).

Looking first at the results from the series based on Diet A, it will be seen from Table 10 that both males and females when on the food of the lowest calcium content averaged lower

		Males		Females	
Age	*Calcium in food (Percent)*	*Series based on Diet A Ca in body (Percent)*	*Series based on Diet B Ca in body (Percent)*	*Series based on Diet A Ca in body (Percent)*	*Series based on Diet B Ca in body (Percent)*
28 days	0.19–0.2	0.715 ± 0.007 [a]	... [a]	0.748 ± 0.006 [a]	... [a]
	0.35		0.740 ± 0.003		0.750 ± 0.004
	0.48	0.956 ± 0.004	0.799 ± 0.004	0.949 ± 0.007	0.821 ± 0.005
	0.64	1.022 ± 0.009	0.854 ± 0.005	1.052 ± 0.012	0.876 ± 0.005
	0.8	0.677 ± 0.005		0.676 ± 0.006	
60 days	0.19–0.2				
	0.35		0.790 ± 0.005		0.901 ± 0.007
	0.48	0.998 ± 0.006	0.866 ± 0.006	1.049 ± 0.006	0.980 ± 0.004
	0.64	1.046 ± 0.010	0.934 ± 0.008	1.105 ± 0.007	1.040 ± 0.006
	0.8	0.726 ± 0.006		0.830 ± 0.008	
90 days	0.19–0.2				
	0.35		0.899 ± 0.005		1.066 ± 0.005
	0.48	1.025 ± 0.005	0.968 ± 0.008	1.163 ± 0.008	1.126 ± 0.006
	0.64	1.061 ± 0.004	1.000 ± 0.007	1.237 ± 0.010	1.187 ± 0.007
	0.8	1.023 ± 0.010		1.195 ± 0.022	
180 days	0.19–0.2				
	0.35		1.032 ± 0.004		1.238 ± 0.005
	0.48	1.112 ± 0.006	1.065 ± 0.007	1.305 ± 0.020	1.311 ± 0.011
	0.64	1.115 ± 0.006	1.098 ± 0.007	1.382 ± 0.013	1.338 ± 0.019
	0.8				

percentages of body calcium at 60 days than at 28 days, because, while the calcium balances had been positive, the gains in body calcium had not kept pace with the gains in body weight during this period; while those on higher levels of calcium intake did increase their percentages of body calcium during this same age period. Yet even on a diet whose calcium content was three times that of Diet A, the percentage of calcium in the body did not reach its plateau value at any time during growth.

TABLE 11

ILLUSTRATING DEGREE OF ATTAINMENT OF EQUALITY IN
BODY CALCIUM AT MIDDLE AGE AMONG RATS REARED ON
DIFFERENT LEVELS OF CALCIUM INTAKE: MALES AT
ONE YEAR OF AGE

Calcium in Food Percent	Series Based on Diet A		Series Based on Diet B	
	Body Calcium Grams	Body Calcium Percent	Body Calcium Grams	Body Calcium Percent
0.19–0.2	3.467	1.105 ± 0.014 [a]		
0.35			3.756	1.096 ± 0.008 [a]
0.48			3.934	1.136 ± 0.008
0.64	3.906	1.176 ± 0.010	4.046	1.170 ± 0.011
0.8	4.022	1.178 ± 0.009		

[a] The precision measure following the ± sign is in each case the probable error of the mean.

From Table 11 it would appear that on the threefold and fourfold levels this was just about attained at middle age; but even then was not attained by the animals on Diet A itself although this diet is adequate according to current standards.

At a given age and calcium-intake level, a comparison either of the males or of the females shows (Tables 9–11) that the body weight is higher for those with dietaries based on Diet B than for those whose dietaries are based on Diet A, and so the former, although they are superior animals and have *larger amounts* of body calcium, have *slightly lower percent-*

ages. Calcium is, of course, not the only factor which contributes to superior nutritional well-being. Note, however, that the animals of each sex, at each age, and with each basal ration studied, always averaged higher percentages of body calcium the higher the percentage of calcium in their food. That this higher percentage of body calcium is an advantage is sometimes superficially challenged as "debatable"; but those who have studied it most carefully are most clearly of the opinion that it is an advantage. I am especially indebted to Dr. P. C. Jeans and to the late Dr. T. Wingate Todd for cogent personal expressions of the view based on their respective researches that this extra body calcium certainly may, and probably will, be a very important asset in the vicissitudes of human life.

Moreover, now that the studies with human beings and those with laboratory animals have been so fully and critically compared, we have what seem to be ample arrays of evidence with which to safeguard our interpretations. For we can view the very extensive findings of direct human experience in the light of the data of fully controlled animal experimentation extending over complete life histories and even successive generations, with chemical analyses of many representative cases at intervals in the life cycle.

Inasmuch as the calcium compounds of the body are so sparingly soluble that less than one percent of the calcium in the body serves to saturate (in the physiological sense, at least) the blood and soft tissues, all the rest being laid up as relatively insoluble bone (and tooth) mineral, it may at first thought appear strange that the acquisition of more calcium can make any important difference to our bodily well-being.

The bone trabeculae are undoubtedly important in this connection. Individually these are delicate growths of bone mineral projecting into the cavity of the bone from the inner surfaces of its porous ends. They are suggestive of the crystal growths often demonstrated in the teaching of chemistry, or of spicules growing from all directions toward the center of the marrow cavity of the bone, or of stalactites and stalagmites similarly growing toward the center of a cave from its sides as well as its roof and floor. Collectively, when well developed, they constitute a closely interlocking criss-cross meshwork like an unusually abundant multiplication of steel bracing rods to aid in supporting and stabilizing a bridge. Yet we should remember that it is not a structure built once for all, but rather it is the changeable total of individual crystals, each of which may increase or decrease by the deposition upon it of calcium salt from the blood or the dissolving away from it of calcium salt by the blood.

This reversible development of bone trabeculae has been cited as an excellent example of *functional hypertrophy:* something which is not of constant occurrence, and so may be regarded as an overgrowth or overdevelopment, yet which functions physiologically to the advantage of the organism that possesses it. In fact, a rich trabecular development functions in two ways: (1) like the multiplication of supplementary rods and braces in a bridge, it affords a maximum of strengthening and stabilizing *mechanical effect* from a relatively small amount of extra material; and (2) by virtue of the large surface which the meshwork arrangement of this extra material exposes, it functions with great efficiency as a *chemical regulator of the calcium (calcium-ion) content of the blood.*

As yet it does not seem to be at all certain whether the so-called senile forms of arthritis and osteomalacia developing in people who have lived for years upon calcium-poor dietaries are attributable to the gradual and cumulative deformation of the ends of the bones for lack of the mechanical support and stabilization of abundant trabecular development; though it is now clear that liberal calcium intake does contribute, in this way among other ways, to the attainment and maintenance of that combination of "chemical maturation" with "youthful condition" of tissue which so largely constitutes and characterizes the prime of life.

It is now quite clearly certain that a rich development of bone trabeculae performs with high efficiency a regulatory function in maintaining that stability of calcium (or calcium-ion) content of blood which is of far-reaching importance both to the immediate well-being of all parts of the body and to the life history of the individual and the family.

It is to be remembered that at and near the ends of the bones their walls are porous and vascular so that the blood circulates freely into and through this part of the bone marrow, in fact a part of the process of blood-making and remaking occurs here. It has also been shown by balance experiments and hematological studies that an equable calcium supply is favorable to the utilization of iron and the most efficient production and regeneration of blood of optimal hemoglobin content. In the light of these facts it is clear that an abundant multiplication and extension of trabeculae, by offering a large surface of bone mineral in immediate and effective contact with the blood at this scene of special physical and chemical activity, has high potentiality for the favorable regulation of the direct environment of a part of the blood-making process and also for the stabilization of the calcium content of the blood as it passes on to its circulation through all parts of the body.

Just before his untimely death, Todd added what may well prove to be an important extension to the story of the bone trabeculae. In his contribution to Cowdry's *Problems of Ageing* (Williams and Wilkins, 1939) he described what appears in skillfully made roentgenograms as "a gray sheen of labile mineral," which in a "well mineralized" bone lies in the interstices between the definite trabecular structures and presumably consists of chemically similar or identical calcium-salt mineral in a semidispersed (though anatomically localized) form presenting still larger surface in proportion to its mass than does the calcium mineral of the fully formed trabeculae themselves. Obviously this addition to our knowledge of the trabeculae accentuates the concept of the efficiency of well-mineralized bone as a regulator of the calcium content of the blood, which in turn we know to be of great importance to all parts of the body and to a wide variety of its functions. What-ever may prove to be the exact significance of this "gray sheen," the general relationships of highly developed trabeculae to liberal intake of food calcium on the one hand and on the other hand to regulation of the calcium and calcium-ion con-centration in the blood are now clear and well established.

Within wide limits, the higher the calcium content of the food, the greater the development of the bone trabeculae and the more promptly is any fluctuation of blood calcium re-stored to the optimal level. This alone would suffice to give an important place to liberal calcium intake among the nutri-tional factors by means of which it is possible to build higher health and longer life.

Experiments with Different Levels of Riboflavin Intake

Largely because riboflavin, then called vitamin B_2 or G, was one of the factors in the difference between the above-noted Diets A and B, experiments with different levels of riboflavin in the food have been carried out in much the same way as those with calcium described in the preceding section. Be-cause of the general similarity of purpose and plan and of the viewpoint from which the data are interpreted, the results of the riboflavin studies may here be summarized very briefly, yet in an adequately critical attitude of mind. It is found with riboflavin as with calcium that there is a wide zone between, on the one hand, the level of minimal adequacy for ostensible health and the perpetuation of the family, and, on the other hand, the level of riboflavin content of food which induces optimal nutritional well-being in long-term experiments with dietaries good also in other respects.

From the immediate viewpoint of the experiments as de-signed to account in explicit chemical terms for the difference in nutritive value between Diets A and B, the findings indi-cate that riboflavin played a larger part than protein and a lesser part than calcium in the superiority of Diet B to Diet A. However, the fact that it here had a less prominent place in the picture than did calcium is doubtless chiefly attributable to the fact that Diet A is already rather richer in riboflavin than is the average American dietary. In the problem of bringing the average American dietary or the total American food supply up to optimal nutritive value in all respects, riboflavin should probably be regarded as coördinate with calcium, when both are regarded as points at which investment in more liberal intake brings good returns in improved nutritional well-being.

Particularly interesting among the results of the successive-generation feeding experiments with different levels of riboflavin intake was the finding that when step-wise enrichments of the dietary had brought it to the "plateau of optimal response" as regards the life histories of original experimental animals, further enrichment still seemed to confer further benefit upon their offspring; and among the points of superiority shown by the young of the highest-riboflavin families was increased ability to endure shortage not only of riboflavin but also of thiamin. Let us consider the significance of this from the viewpoint of our present discussion.

While our predominant working concept is that of the specific function of, and need for, each nutritionally essential factor (chemical element or compound as the case may be)—and in this connection we have widely useful "yardsticks" of requirements and recommended allowances of specific nutrients—yet also if our picture is to be fully valid it must recognize the supplementary concept or principle of nutritional flexibility. This latter principle may enter into the problem of the attainment and maintenance of optimal nutritional well-being in one or more of at least three ways. There may be an apparently direct elasticity by which the body adapts itself to a higher or lower intake. Also, the functions of nutrients, while specific in their more obvious aspects, may involve interrelationships of two kinds. On the one hand, while each of two processes is specific, both processes may yet be, in chemical parlance, "linked," or (in mechanical or electrical analogy) may function "in series" so that the efficacy of one is dependent upon the adequacy of the other, as illustrated in findings with certain vitamin-enzymes by Elvehjem and co-workers at Wisconsin. Or, on the other hand, two things, each of them essential and working in the main at its specific function, may yet be able to "spare," or in some degree substitute for, each other, as illustrated in thé long-known protein-sparing power of carbohydrates and fats, and in the much more recent discovery by Ellis and Zmachinsky at Columbia that riboflavin can in some way "spare" thiamin. This latter fact can be conceived in either (or both) of two ways: (1) These two vitamins form body enzymes which, while different, yet work in a sufficiently parallel way so it is conceivable that one of them can, at least in some limited degree, substitute for the other, the material which is undergoing metabolism (or at least some of it) being drawn through whichever path or channel is most fully and effectively provided with the enzymes which catalyze its transformation. (2) The two substances riboflavin and thiamin have enough in common in their molecular structure so that we can conceive that in their own break-down in the body they may yield some common product, in which case, according to the universally accepted "mass action principle" or "concentration law," the utilization of the abundantly supplied riboflavin would in the literal sense spare and conserve the body's relatively more limited store of thiamin.

However the still intricate problems of interrelationships among the B vitamins may turn out, there is already clear evidence that riboflavin is one of the chemical factors or specific nutrients through liberal use of which passable health may be built to higher levels.

Experiments with Dietaries of Differing Vitamin A Values

Many experiments with dietaries of differing vitamin A values have been made in different laboratories, including those of the Columbia Department of Chemistry where investigations in this field have been carried on in part with a view to the chemical elucidation of the nutritional difference between Diets A and B, and in part as independent researches.

Early among the latter was the work of Dr. F. L. MacLeod which showed that differences of vitamin A value in the dietary may show practically no effects over a large part of the life history, including all of the period of most rapid growth, and yet may be of very great importance to middle and later life and to the perpetuation of the family.

One reason for the relatively slow appearance of the important ultimate effects of different vitamin A intakes is the large capacity of the body for storage of vitamin A even in infancy; this was shown by the experiments of Dr. L. B. Storms and of Dr. M. L. Cammack, which closely followed those of Dr. F. L. MacLeod in the Columbia laboratory. Then the later experiments of Dr. H. L. Campbell in the same laboratory indicated that the constructive benefit of the more liberal intakes of vitamin A may be in some way inherently or specifically greater at the more mature ages. This is now being studied further at Columbia.

Attacking the same general problem in a different way, Dr. E. L. Batchelder fed laboratory animals through two generations upon diets alike in other respects but representing four levels of vitamin A intake: that of approximately minimal adequacy, and two, four, and eight times this minimum. The original experimental animals, placed on these different diets at the end of infancy, showed successively increased benefit from the successive increments of vitamin A, up to about fourfold the minimal adequate level. Here they seemed to have reached their plateau of optimal response, but the vigor of their offspring was still benefited by a further rise in the vitamin A value of the family dietary. This now seems to be very largely a question of what the mother's food permits her to do for her offspring. On one occasion when Dr. Batchelder's findings were being discussed, a geneticist asked: "What do you mean by benefit to the offspring; have you any improvement there that is transmissible through the father?" And when told that we had no expectation of changing the germ plasm by improving the diet, he asked: "Then does the higher vitamin A intake do anything more than make the mother a better producer of milk?" The geneticist's two questions were doubtless both challenges to debate—first one extreme view, and then another at the opposite extreme.

Either a geneticist or a nutritionist, seriously seeking the strongest scientific probability in the interpretation of Batchelder's data, would doubtless look for it on ground intermediate between the implications of these two questions. Perhaps a higher vitamin A value of diet may be conducive to a better production of milk, but it is not reasonably probable to suppose that this alone would tell the whole story, or the part of the story most significant for the largest number of human beings.

The stronger probability is that in addition to whatever extra the mother may have been enabled to do through her mammary glands, she has also been enabled, by the higher vitamin A value of her diet, to maintain a superior internal environment which is advantageous to the prenatal as well as postnatal nourishment of the young. Moreover, this same superiority of internal environment is presumably not specifically related to reproduction and lactation but more probably is also a factor in the broader and life-long project of the nutritional improvement of the norm.

"Nature and Nurture"

In the days of Darwin, science awoke to the importance of heredity; and the wave of interest therein aroused has been in full flood for two generations. True, the present-day geneticist may make such use of mutation in his explanations as to seem to have a view rather different from that which Darwin held as to how things happen, yet inheritance of advantageous characteristics, or original chromosomal endowment, still seems "the best explanation" of a superior life history. Hence a superiority for which a genetic explanation can be given is thereby dignified and made more impressive, and conversely a superiority not genetically accounted for is apt to be ignored or belittled, "explained away," as in the second question of the geneticist's discussion of the benefit of vitamin A mentioned in the preceding section.

As a favorable life history was thus explained genetically, so an unfavorable one could be explained as due either to bad luck in one's chromosomal endowment or to the pathogens in one's environment. For through the work of Pasteur and his contemporaries, the great importance to human life of the bacteriology and sanitation of the body's surroundings was made clear; and for a generation students of health have naturally and properly concentrated their attention upon the triumphant use of bacteriology and sanitation in the control of infectious diseases.

Thus the impression grew very strong that the two great factors in having high health were, first, good luck in one's original chromosomal endowment, and then, avoidance of injury from one's surroundings. Each of these two concepts came to be so greatly appreciated that together they seemed to explain all the major differences in human life histories. It came to be regarded as a truism that "heredity and environment determine our lives"; and while *by definition* nutrition was admitted as an environmental factor, yet *in ordinary thinking environment meant surroundings* to such an extent that the importance of the internal environment was not sufficiently realized to lead to any effective challenging of the dogma that it was *fixed* by nature.

Hence it became a habit, of scientists as well as of the public, to attribute to heredity or nature or chromosomal endowment any difference for which the surroundings did not afford an explanation. It was realized, of course, that the bodily "mechanism" given us by nature must be nourished; but the nourishing of it was too mechanically and passively conceived.

Thus even the concept of deficiency diseases had up-hill

work to win recognition because it was thought that for every disease there must be some *materies morbi*. It seemed too fanciful to suppose that a disease could be "due to something you don't have." And while the evidence has now become so clear that one can no longer doubt that several previously baffling diseases are due to shortages of normal nutrients, there is still a strong tendency to admit only the negative or corrective without the corresponding positive or constructive aspect of the new concept into our fully effective thinking.

An idea is more readily and effectively adopted into our thinking when it solves a dilemma than when it involves unlearning something that we supposed we knew. So the *corrective* aspect of the newer knowledge of nutrition, which furnishes explanations for the recent dilemmas of baffling diseases and failures of laboratory animals to grow on mixtures of previously known nutrients, has found quicker and fuller acceptance than has yet been effectively accorded to the *constructive* aspect, which relates nutrition to superior health which we thought was attributable to heredity, or original good luck in chromosomal endowment, or to "natural constitution."

Hopkins was trying, by means of persuasive understatement, to educate both biological and chemical contemporaries to the constructive as well as the remedial aspect of the new knowledge when he said that "Nurture can assist Nature" to a larger extent than the Science of the past generation or two has thought.

Marked skepticism as to the constructive potentialities of nutrition should now be recognized as a too-fatalistic hangover from the period before science had its present research methods and objective evidence accumulated under well-controlled conditions.

Todd (1935) wrote that "The adult physical pattern is the outcome of growth along lines determined by heredity but enhanced, dwarfed, warped or mutilated in its expression by the influence of environment in the adventures of life." This was explicit recognition, arising from anatomical research, that one's hereditary expectation can be "enhanced" by environmental conditions just as truly as also it can be "dwarfed, warped, or mutilated." That hereditary expectation can be enhanced by nutrition is perhaps a clearer expression of the same thought expressed by Hopkins when he said that nurture can "assist" nature to an extent not previously apprehended by science.

Mendel and Hubbell (1935) found that, under the favorable nutritional conditions maintained in their rat colony, there was increased growth and development which gradually progressed through successive generations in a way which could not have been predicted from heredity; and the more expeditious growth and development was a real advantage as shown by the subsequent adult vitality or stamina of the same animals. Similarly in the above-described improvement of an already adequate diet, Sherman and Campbell (1924, 1930, 1935) likewise induced higher health. In this latter case the objective evidence also showed a well marked increase in the length of adult life and a still larger extension of the period of the prime or of full adult capacity—the period between the attainment of maturity and the onset of senility.

Perhaps the evidence is more clearly expressed by the statement that the moderately more rapid growth and development was both accompanied and followed by superior stamina and vitality, the increased length of life being consequent upon the life having been lived at higher health levels throughout. Thus one can properly speak of higher health and longer life being built in the same individual by the same nutritional improvement.

Or adapting Streeter's analogy of life as a ballistic curve or arc, one may liken superior nutrition to a reinforcement ("enhancement," in Todd's phrase) of the original life impulse somewhat as, with a properly aimed cannon, superior gunpowder throws the projectile both higher and farther.

The slowness of people to grasp the full significance of the nutritional improvement of life is apparently not so much due to anything inherently difficult to understand about the evidence furnished by nutritional research, but rather to the fact that many people are not quite open-minded in this direction, because their minds are preoccupied by the impression that the lengths of our lives and the health-levels upon which we can live them are predestined by our heredity or the chromosomal endowment received at conception. Thus a part of the truth was too fatalistically accepted as if it were the whole, and now that another important part has been found, it at first may not seem to fit in. There has been too strong a feeling of contentment in the previously held view. In holding the view that "heredity is the best explanation," one could have the double satisfaction of being in the mode and of explaining the differences of health level among passably normal people in a way that did not call upon one to do anything about it: too largely one's "constitution" appeared to have been predestined from his conception. How much of what is called "constitution" is in fact due to the chromosomal endowment and how much is due to the body's internal environment which the individual may (and often unconsciously does) influence for better or worse through his everyday food habits should now be regarded as an open question. It is also an important question; and one just now opened to objective, well-controlled investigation by the newly developed methods of nutritional research, and upon which much light may perhaps be thrown by the nutritional findings now being currently reported.

It may well be that thorough and long-term use of the guidance which the science of nutrition now offers might greatly ameliorate much of what now passes for "constitutional inadequacy" and, functioning constructively as well as correctively, might advance the level of health of many only passably healthy people to the status of the more fortunate people supposedly born with better constitutions.

Dr. W. F. Dove, research biologist in the Agricultural Experiment Station of the University of Maine, has thrown a very interesting and suggestive light upon the general question of what is innate and what is nutritionally improvable, by his emphasis upon nutritional instinct as an innate characteristic subject to individual variation. The fact that food supply may be responsible for what are often wrongly held to be differences of inborn potentiality had been emphasized by Hopkins, speaking for both biological and chemical science as president of the Royal Society. Dr. Dove points out that the "innately superior" animal in a farm flock or herd thrives better in a given environment than do his fellows, largely because he is born with instincts which lead him to make a better-than-average use of what the nutritional environment affords, in other words, his inborn qualities include a superiority of nutritive instinct as to which of the available foods to eat, and in what relative amounts. If, then, the other animals are fed according to his example, they develop better than they would without this scientific guidance of their nutrition.

What Dove here shows feasible in controlled animal feeding is fairly analogous to what is done when children are given extra milk, as in Mann's investigation and the British Milk-in-Schools Scheme, regardless of whether they would have sought it on their own initiative; but with the result that when they have the extra milk they exceed their previous rate of development, attaining as Fletcher explicitly states to higher rates of mental as well as physical growth.

Nutritional improvement of the length of adult life is a concept which most people are slower to grasp than the improvement of growth and development. Doubtless this is largely because prior to the recent advances in nutrition there had been great advances in other aspects of public health, which in the course of two or three generations had greatly increased the life expectation of the infant by the reduction of the death rates of the early ages, but had not reduced the death rates of the latter part of the life cycle, and so had not improved the life expectation of the adult. Now, however, research has definitely and objectively shown how nutrition can reduce the death rates both of children and of adults, and so can improve the *adult* life expectation.

Even more strongly should we emphasize the fact that the lives which are made longer by such improvements in nutrition as we are considering *get to be longer because they are lived on a higher level of health.* The extra years, whatever their number, are not added to the period of old age, but are inserted in the period of the prime, at its apex if you choose. Or in another analogy this effect of a superior nutrition means that the *plateau* of the prime of life is both higher and longer.

Those who think it was already settled that the only way to have longer life is to select a longer-lived ancestry have often misinterpreted the work of the late Raymond Pearl into an appearance of support for their view. But Pearl's own view was simply that in his early studies he had correlated longevity with heredity and not with other factors. In his later work he did correlate the lengths of men's lives with certain differences in their daily habits, all previously supposed good. So Pearl's published research as a whole supports the view here held: that both what a man is born with and what he does himself are significant in determining the length of his life. This is also the view of Dublin and his coworkers. While neither they nor Pearl have specifically investigated nutrition as distinguished from other environmental factors, they have published findings which make it no longer true that the statistical studies of human longevity point only to hereditary factors.

Conscious chemical control and improvement of one's internal environment through one's nutritional habits is now offered us by science as a workable proposition, a practicable opportunity.

Being internal and having to do rather with the efficiency of coördination of the life processes than with the great accentuation of any one of them, such chemical improvement through nutrition does not show itself dramatically as a mutation may, and as do some of the exploits of endocrinology; but it may be no less far-reaching because it is working in harmony with human evolution. Moreover, by its very success this newer chemistry of nutrition is rapidly outgrowing the stage of easily dramatized advances such as are reducible to a few specific words. The question, What particular thing does such nutritional improvement do? hardly admits of a simpler or more specific answer than that it offers us the option of a more liberal term of years of more efficient life with which to do what we will.

APPENDIX

RECOMMENDED DIETARY ALLOWANCES *

Food and Nutrition Board, National Research Council

	Calories	Protein grams	Calcium grams	Iron mg.	Vitamin A*** I.U.	Thiamin (B_1) mg.**	Riboflavin mg.	Niacin (Nicotinic acid) mg.	Ascorbic acid mg.**	Vitamin D I.U.
Man (70 Kg.)										
Sedentary	2500	......	......	....	......	1.5	2.2	15	.....	.....
Moderately active	3000	70	0.8	12	5000	1.8	2.7	18	75	†††
Very active	4500	......	......	....	......	2.3	3.3	23	.....	.....
Woman (56 Kg.)										
Sedentary	2100	......	......	....	......	1.2	1.8	12	.....	.....
Moderately active	2500	60	0.8	12	5000	1.5	2.2	15	70	†††
Very active	3000	......	......	....	......	1.8	2.7	18	.....	.....
Pregnancy (latter half)	2500	85	1.5	15	6000	1.8	2.5	18	100	400 to 800
Lactation	3000	100	2.0	15	8000	2.3	3.0	23	150	400 to 800
Children up to 12 years:										
Under 1 year †	100/Kg.	3 to 4/Kg.	1.0	6	1500	0.4	0.6	4	30	400 to 800
1–3 years ††	1200	40	1.0	7	2000	0.6	0.9	6	35	†††
4–6 years	1600	50	1.0	8	2500	0.8	1.2	8	50	.....
7–9 years	2000	60	1.0	10	3500	1.0	1.5	10	60	.....
10–12 years	2500	70	1.0	12	4500	1.2	1.8	12	75	.....

Children over 12 years:

Girls, 13–15 years	2800	80	1.3	15	5000	1.4	2.0	14	80	†††
16–20 years	2400	75	1.0	15	5000	1.2	1.8	12	80	....
Boys, 13–15 years	3200	85	1.4	15	5000	1.9	2.4	16	90	†††
16–20 years	3800	100	1.4	15	6000	2.3	3.0	20	100	....

• Tentative goal toward which to aim in planning practical dietaries; can be met by a good diet of natural foods. Such a diet will also provide other minerals and vitamins, the requirements for which are less well known.

•• 1 mg. thiamin equals 333 I.U.; 1 mg. ascorbic acid equals 20 I.U.

••• Requirements may be less if provided as vitamin A; greater if provided chiefly as the pro-vitamin carotene.

† Needs of infants increase from month to month. The amounts given are for approximately 6–8 months. The amounts of protein and calcium needed are less if derived from breast milk.

†† Allowances are based on needs for the middle year in each group (as 2, 5, 8, etc.) and for moderate activity.

††† Vitamin D is undoubtedly necessary for older children and adults. When not available from sunshine, it should be provided probably up to the minimum amounts recommended for infants.

Further Recommendations, Adopted 1942:

The requirement for *iodine* is small; probably about 0.002 to 0.004 milligram a day for each kilogram of body weight. This amounts to about 0.15 to 0.30 milligram daily for the adult. This need is easily met by the regular use of iodized salt; its use is especially important in adolescence and pregnancy.

The requirement for *copper* for adults is in the neighborhood of 1.0 to 2.0 milligrams a day. Infants and children require approximately 0.05 per kilogram of body weight. The requirement for copper is approximately one-tenth of that for iron.

The requirement for *vitamin K* is usually satisfied by any good diet. Special consideration needs to be given to newborn infants. Physicians commonly give vitamin K either to the mother during pregnancy or to the infant immediately after birth.

SELECTED BIBLIOGRAPHY

Armsby, H. P. 1913. Food as body fuel. Bulletin 126 of the Pennsylvania Agricultural Experiment Station (State College, Pennsylvania).

Arnold, A., and C. A. Elvehjem. 1939. Processing and thiamin. Food Research 4: 547–553.

Atwater, W. O., and F. G. Benedict. 1905. A respiration calorimeter with appliances for the direct determination of oxygen. Carnegie Institution of Washington, Publication 42.

Aughey, E., and E. P. Daniel. 1940. Effect of cooking upon the thiamin content of foods. J. Nutrition 19: 285–296.

Batchelder, E. L. 1934. Nutritional significance of vitamin A throughout the life cycle. Am. J. Physiol. 109: 430–435.

—— 1934. Vitamin C in apples. J. Nutrition 7: 647–655.

Bauer, W., J. C. Aub, and F. Albright. 1929. Bone trabeculae as a readily available reserve supply of calcium. J. Exptl. Med. 49: 145–161.

Benedict, F. G. 1918. A portable respiration apparatus for clinical use. Boston Med. Surg. J. 178: 667–678.

—— 1928. Basal metabolism: the modern measure of vital activity. The Scientific Monthly 27: 5–27.

Benedict, F. G., and C. G. Benedict. 1933. Mental effort in relation to gaseous exchange, heart rate, and mechanics of respiration. Carnegie Institution of Washington, Publication 446.

Bessey, O. A. 1938. Vitamin G and synthetic riboflavin. J. Nutrition 15: 11–15.

Bessey, O. A., and R. L. White. 1942. The ascorbic acid requirements of children. J. Nutrition 23: 195–204.

Bessey, O. A., and S. B. Wolbach. 1938, 1939. Vitamin A: physiology and pathology. J. Am. Med. Assoc. 110: 2072–2080. Reprinted as Chapter II of "The Vitamins, 1939." Chicago: American Medical Association.

Bessey, O. A., and S. B. Wolbach. 1939. Vascularization of the cornea of the rat in riboflavin deficiency. J. Exptl. Med. 69: 1–12.

Best, C. H., and J. H. Ridout. 1939. Choline as a dietary factor. Ann. Rev. of Biochem. 8: 349–370.

Bisbey, B., et al. 1934. The vitamin A and D activity of egg yolks of different color concentrations. Missouri Agricultural Experiment Station Research, Bull. 205.

Blatherwick, N. R., and M. L. Long. 1922. Some effects of drinking large amounts of orange juice and sour milk. J. Biol. Chem. 53: 103–109.

Bloch, C. E. 1924, 1931. (Effects of shortage of vitamin A in children.) Am. J. Diseases Children 27: 139–148; 42: 263–278.

Blunt, K., and R. Cowan. 1930. Ultraviolet Light and Vitamin D in Nutrition. Chicago: University of Chicago Press.

Booher, L. E. 1939. Chemical aspects of riboflavin. Chapter XIII of "The Vitamins, 1939." Chicago: American Medical Association.

Booher, L. E., and R. A. Hartzler. 1939. (Thiamin contents of certain foods.) U.S. Dept. Agriculture, Tech. Bull. 707.

Borsook, H. 1940. The oxidation-reduction potential of coenzyme I. J. Biol. Chem. 133: 629–630.

Boudreau, F. G. 1937. The international campaign for better nutrition. Milbank Memorial Fund Quarterly 15: 103–120.

Boudreau, F. G., and H. D. Kruse. 1939. Malnutrition: A challenge and an opportunity. Am. J. Public Health 29: 427–433.

Boudreau, F. G., and others. 1942. The food and nutrition of industrial workers in wartime. National Research Council, Reprint and Circular Series, No. 110.

Boynton, L. C., and W. L. Bradford. 1931. Effect of vitamins A and D on resistance to infection. J. Nutrition 4: 323–329.

Briwa, K. E., and H. C. Sherman. 1941. The calcium content of the normal growing body at a given age. J. Nutrition 21: 155–162.

Burnet, E., and W. R. Aykroyd. 1935. Nutrition and public health. League of Nations Health Organization Quarterly Bull. 4: 326–474.

Caldwell, M. L., L. E. Booher, and H. C. Sherman. 1931. Crystalline amylase. Science 74: 37.

Campbell, H. L. 1928, 1931. Growth, reproduction, and longevity of experimental animals as research criteria in the chemistry of nutrition. Dissertation, Columbia University; and J. Am. Dietet. Assoc. 7: 81–94.

Campbell, H. L., O. A. Bessey, and H. C. Sherman. 1935. Adult rats of low calcium content. J. Biol. Chem. 110: 703–706.

Campbell, H. L., and H. C. Sherman. 1938. Nutritional effects of the addition of meat and green vegetable to a wheat-and-milk diet. J. Nutrition 15: 603–612.

Cannon, W. B. 1939. The Wisdom of the Body, Revised Edition. New York: W. W. Norton and Company.

Carpenter, T. M. 1915. A comparison of methods for determining the respiratory exchange in man. Carnegie Institution of Washington, Publication 216.

Chenowith, L. B. 1937. Increase in height and weight and decrease in age of college freshmen. J. Am. Med. Assoc. 108: 354–356.

Chittenden, R. H. 1907. The Nutrition of Man. New York: Stokes.

Clausen, S. W. 1934. The influence of nutrition upon resistance to infection. Physiol. Rev. 14: 309–350.

Conant, J. B. 1939. The Chemistry of Organic Compounds. Revised edition. New York: The Macmillan Company.

Coward, K. H. 1937. The Biological Standardization of Vitamins. London: Balliere, Tindall, and Cox.

Cowgill, G. R. 1939. The physiology of vitamin B_1. Chapter VIII of "The Vitamins, 1939." Chicago: American Medical Association.

Dam, H. 1940. Fat-soluble vitamins. Ann. Rev. of Biochem. 9: 353–382.

Day, P. L., W. J. Darby, and K. W. Cosgrove. 1938. The arrest of nutritional cataract by the use of riboflavin. J. Nutrition 15: 83–90.

De Kleine, W. 1942. Control of pellagra. Southern Med. J. 35: 992–996.

Deuel. H. J., Jr., and others. 1928. A study of the nitrogen minimum. J. Biol. Chem. 76: 391–406, 407–414.

Doisy, E. A., S. B. Binkley, and S. A. Thayer. 1941. Vitamin K. Chem. Rev. 28: 477–517.

Dove, W. F. 1935. A study of individuality in the nutritive instincts and of the causes and effects of variations in the selection of food. Am. Naturalist 69: 469–544.

——— 1939. The needs of superior individuals as guides to group ascendance. J. Heredity 30: 157–163.

Drummond, J. C. 1938. The fat-soluble vitamins. Ann. Rev. of Biochem. 7: 335–340.

Du Bois, E. F. 1936. Basal Metabolism in Health and Disease. 3d edition. New York: Lea and Febiger.

Dutcher, R. A., P. L. Harris, E. R. Hartzler, and N. B. Guerrant. 1934. The assimilation of carotene and vitamin A in the presence of mineral oil. J. Nutrition 8: 269–283.

du Vigneaud, V., J. P. Chandler, A. W. Moyer, and D. M. Keppel. 1939. The effect of choline on the ability of homocystine to replace methionine in the diet. J. Biol. Chem. 131: 57–76.

Eddy, W. H., and G. Dalldorf. 1941. The Avitaminoses. 2d edition. Baltimore: Williams and Wilkins.

Eliot, M. M., and E. B. Jackson. 1933. Bone development of infants and young children in Puerto Rico. Am. J. Diseases Children 46: 1237–1262.

Eliot, M. M., E. M. Nelson, S. P. Souther, and M. K. Cary. 1932. The value of salmon oil in the treatment of infantile rickets. J. Am. Med. Assoc. 99: 1075–1082.

Ellis, L. N., A. Zmachinsky, and H. C. Sherman. 1943. Effects of liberal intakes of riboflavin. J. Nutrition (25: 153–160).

Elvehjem, C. A. 1935. The biological significance of copper and its relation to iron metabolism. Physiol. Rev. 15: 471–507.

—— 1939. Nicotinic acid in nutrition. Ann. Internal Med. 13: 225–231.

—— 1942. The water-soluble vitamins. J. Am. Med. Assoc. 120: 1388–1397.

Elvehjem, C. A., W. H. Peterson, and D. R. Mendenhall. 1933. Hemoglobin content of the blood of infants. Am. J. Diseases Children 46: 105–112.

Fairbanks, B. W. 1938. Improving the nutritive value of bread by the addition of dry milk solids. Cereal Chemistry 15: 169–180.

Farrar, G. E., Jr., and S. M. Goldhamer. 1935. The iron requirement of the normal human adult. J. Nutrition 10: 241–254.

Fincke, M. L. 1941. The utilization of the calcium of broccoli and cauliflower. J. Nutrition 22: 477–482.

Fincke, M. L., and H. C. Sherman. 1935. Availability of calcium from some typical foods. J. Biol. Chem. 110: 421–428.

Folin, O. 1905. A theory of protein metabolism. Am. J. Physiol. 13: 117–138.

Folin, O., and J. L. Morris. 1913. The normal protein metabolism of the rat. J. Biol. Chem. 14: 509–515.

Forbes, E. B., and M. H. Keith. 1914. A review of the literature of phosphorus compounds in animal metabolism. Ohio Agricultural Experiment Station, Tech. Bull. 5.

Frey, C. N., A. S. Schultz, and L. Atkin. 1940. Improving the nutritive properties of bread. Proc. Food Conf. Inst. Food Technologists, June 16–19, 1940, pp. 275–278.

Goldberger, J., G. A. Wheeler, R. D. Lillie, and L. M. Rogers.

1926. (Pellagra-preventing vitamin.) U.S. Public Health Repts. 41: 297–318.

Greenberg, D. M., and W. W. Campbell. 1940. Studies in mineral metabolism with the aid of induced radioactive isotopes. IV. Manganese. Proc. National Acad. Sci. 26: 448–452.

Griffith, W. H., and N. J. Wade. 1940. Choline metabolism. II. The interrelationship of choline, cystine, and methionine in the occurrence and prevention of hemorrhagic degeneration in young rats. J. Biol. Chem. 132: 627–637.

Gunderson, F. L., and H. Steenbock. 1932. Is the vitamin B content of milk under physiological control? J. Nutrition 5: 199–211.

György, P. 1935. Investigations on the vitamin B_2 complex. I–III. Biochem. J. 29: 741–775.

György, P., F. S. Robscheit-Robbins, and G. H. Whipple. 1938. Riboflavin increases hemoglobin production in the anemic dog. Am. J. Physiol. 122: 154–159.

Hambidge, G. 1934. Your Meals and Your Money. New York: McGraw-Hill Company.

—— 1939. Food and life—a summary. U.S. Dept. Agriculture Yearbook for 1939, pages 3–94.

—— 1939b. Nutrition as a national problem. J. Home Economics 31: 361–364.

Harris, L. J., and P. C. Leong. 1936. The excretion of vitamin B_1 in human urine and its dependence on the dietary intake. Lancet 1936, I: 886–894.

Hart, E. B., and H. Steenbock. 1919. Maintenance and production values of some protein mixtures. J. Biol. Chem. 38: 267–285.

Hart, E. B., H. Steenbock, J. Waddell, and C. A. Elvehjem. 1928. Copper as a supplement to iron for hemoglobin building in the rat. J. Biol. Chem. 77: 797–812.

Hastings, A. B., and C. W. Eisele. 1940. Diurnal variations in the acid-base balance. Proc. Soc. Exptl. Biol. and Med. 43: 308–312.

Hecht, S., and J. Mandelbaum. 1939. The relation between vitamin A and dark adaptation. J. Am. Med. Assoc. 112: 1910–1916.

Henderson, L. J. 1924. The Fitness of the Environment. New York: The Macmillan Company.

Henderson, Y., and H. W. Haggard. 1925. The maximum of human power and its fuel. Am. J. Physiol. 72: 220, 264–282.

Herter, C. A., and A. I. Kendall. 1910. The influence of dietary alterations on the types of intestinal flora. J. Biol. Chem. 7: 203–236.

Hess, A. F. 1920. Scurvy, Past and Present. Philadelphia: J. B. Lippincott.

—— 1929. Rickets, Osteomalacia, and Tetany. New York: Lea and Febiger.

Hitchcock, D. I. 1934. Physical Chemistry for Students of Biology and Medicine. 2d edition. Springfield, Ill.: Charles C. Thomas.

Hogan, A. G. 1938. Riboflavin: physiology and pathology. J. Am. Med. Assoc. 110: 1188–1193.

Hoh, P. W., J. C. Williams, and C. S. Pease. 1934. Possible sources of calcium and phosphorus in the Chinese diet. I. The determination of calcium and phosphorus in a typical Chinese dish containing meat and bone. J. Nutrition 7: 535–546.

Hopkins, F. G. 1931. Nutrition and human welfare. Nutrition Abstracts and Reviews 1: 3–6.

—— 1933. Some chemical aspects of life. Science 78: 219–231.

Howe, P. R., R. L. White, and M. D. Eliot. 1942. The influence of nutrition supervision on dental caries. J. Am. Dental Assoc. 29: 38–43.

Irwin, M. H. 1939. Milk as a food throughout life. Wisconsin Agricultural Experiment Station, Bull. 447.

Jeans, P. C. 1941. Marriott's Infant Nutrition. 3d edition. St. Louis: C. V. Mosby Company.

Jeans, P. C., and G. Stearns. 1938. The human requirement of vitamin D. J. Am. Med. Assoc. 111: 703–711.

Jolliffe, N., J. S. McLester, and H. C. Sherman. 1942. The prevalence of malnutrition. J. Am. Med. Assoc. 118: 944–950.

King, C. G. 1939. The chemistry and physiology of vitamin C. Chapters XVII and XVIII of "The Vitamins, 1939." Chicago: American Medical Association.

—— 1939b. The water-soluble vitamins. Ann. Rev. of Biochem. 8: 371–414.

Kruse, H. D. 1941. Medical evaluation of nutritional status. IV. The ocular manifestations of avitaminosis A, with especial consideration of the detection of early changes by biomicroscopy. Public Health Repts. 56: 1301–1324.

—— 1942. A concept of the deficiency states. Milbank Memorial Fund Quarterly 20: 245–261.

Kruse, H. D., V. P. Sydenstricker, W. H. Sebrell, and H. M. Cleckley. 1940. Ocular manifestations of ariboflavinosis. Public Health Repts. 55: 157–169.

Lanford, C. S. 1939. The effect of orange juice on calcium assimilation. J. Biol. Chem. 130: 87–95.

—— 1942. Studies of liberal citrus intakes. J. Nutrition 23: 409–416.

Lanford, C. S., H. L. Campbell, and H. C. Sherman. 1941. Influence of different nutritional conditions upon the level of attainment in the normal increase of calcium in the growing body. J. Biol. Chem. 137: 627–634.

Leitch, I. 1937, 1938. The determination of the calcium requirement of man. Nutrition Abstracts and Reviews 6: 553–578; 8: 1–2.

Lepkovsky, S. 1940. The water-soluble vitamins. Ann. Rev. of Biochem. 9: 383–422.

Leverton, R. M. 1941. Iron metabolism in human subjects on daily intakes of less than five milligrams. J. Nutrition 21: 617–631.

Lewis, H. B. 1935. The chief sulfur compounds in nutrition. J. Nutrition 10: 99–116.

—— 1942. Protein in nutrition. J. Am. Med. Assoc. (A review.) 120: 198–204.

Lipmann, F. 1937. Metabolism of pyruvic acid and the mechanism of vitamin B_1 action. Skand. Arch. Physiol. 76: 255–272.

Lipschitz, M. A., Van R. Potter, and C. A. Elvehjem. 1938. The relation of vitamin B_1 to cocarboxylase. Biochem. J. 32: 474–484.

Logan, M. A. 1940. Recent advances in the chemistry of calcification. Physiol. Rev. 20: 522–560.

Lund, C. C., and J. H. Crandon. 1941. Human experimental scurvy. J. Am. Med. Assoc. 116: 663–668.

Lusk, G. 1928. The Elements of the Science of Nutrition. 4th edition, Philadelphia: W. B. Saunders.

McCay, C. M., L. A. Maynard, G. Sperling, and L. L. Barnes. 1939. Retarded growth, life span, ultimate body size, and age changes in the albino rat after feeding diets restricted in calories. J. Nutrition 18: 1–13.

McClendon, J. F. 1939. Iodine and the Incidence of Goiter. Minneapolis: University of Minnesota Press.

McCollum, E. V., E. Orent-Keiles, and H. G. Day. 1939. The Newer Knowledge of Nutrition. 5th edition. New York: The Macmillan Company.

McGee, L. C., and J. D. Creger. 1942. Gastrointestinal disease among industrial workers. J. Am. Med. Assoc. 120: 1367–1369.

Mack, P. B., and A. P. Sanders. 1940. The vitamin A status of families in widely different economic levels. Am. J. Med. Sci. 199: 686–697.

MacLeod, F. L., *et al.* 1935. The vitamin A content of five varieties of sweetpotato. J. Agric. Research 50: 181–187.

MacLeod, G. 1924. Studies of the Normal Basal Energy Requirements. Dissertation, Columbia University.

McLester, J. S. 1935. Nutrition and the future of man. J. Am. Med. Assoc. 104: 2144–2147.

—— 1939. Borderline cases of nutritive failure. J. Am. Med. Assoc. 112: 2110–2114.

—— 1942. Handbook of nutrition: 1, Introduction. Published in J. Am. Med. Assoc. 119: 945–947.

Macy, I. G. 1942. Nutrition and Chemical Growth in Childhood. I. Evaluation. Springfield, Ill.: Charles C. Thomas.

Manly, M. L., and W. F. Bale. 1939. The metabolism of inorganic phosphorus of rat bones and teeth as indicated, by the radioactive isotope. J. Biol. Chem. 129: 125–134.

Mann, H. C. C. 1939. Diet during the school age. J. Royal Inst. Public Health 2: 486–514.

Marine, D. 1935. The pathogenesis and prevention of simple or endemic goiter. J. Am. Med. Assoc. 104: 2334–2341.

Maynard, L. A. 1937. Animal Nutrition. New York: McGraw-Hill Pub. Co.

—— 1942. Foods of plant origin. J. Am. Med. Assoc. 120: 692–697.

Mendel, L. B. 1923. Nutrition: the Chemistry of Life. New Haven: Yale University Press.

Mendel, L. B., R. B. Hubbell, and A. J. Wakeman. 1937. The influence of some commonly used salt mixtures upon growth and bone development of the albino rat. J. Nutrition 14: 261–272.

Mitchell, H. H., and T. S. Hamilton. 1929. The Biochemistry of the Amino Acids. New York: Chemical Catalogue Company.

Mitchell, H. S., O. A. Merriam, and E. L. Batchelder. 1939. The vitamin C status of college women. J. Home Economics 30: 645–650.

Moore, T. 1937. The vitamin A reserve of the adult human being in health and disease. Biochem. J. 31: 155–164.

Morgan, A. F., A. Field, and P. F. Nichols. 1931. Effect of drying and sulfuring on vitamin C content of prunes and apricots. J. Agric. Research 42: 35–45.

Murlin, J. R., R. E. Conklin, and M. E. Marsh. 1925. Energy metabolism of normal newborn babies with special reference to the influence of food and crying. Am. J. Diseases Children 29: 1–28.

Murlin, J. R., M. E. Marshall, and C. D. Kochakian. 1941. Digestibility and biological value of whole wheat breads as compared with white bread. J. Nutrition 22: 573–588.

Murphy, E. F. 1941. Ascorbic acid content of onions and observations on its distribution. Food Research 6: 581–594.

—— 1942. The ascorbic acid content of different varieties of Maine-grown tomatoes and cabbages as influenced by locality, season, and stage of maturity. J. Agric. Research 64: 483–502.

National Nutrition Conference. 1942. Proceedings of the Conference held in Washington, May, 1941. Washington: Government Printing Office.

Newburgh, L. H., and M. W. Johnston. 1930. The Exchange of Energy between Man and the Environment. Springfield, Ill.: Charles C. Thomas.

Northrop, J. H. 1939. Crystalline Enzymes. New York: Columbia University Press.

Ochoa, S., and R. J. Rossiter. 1939. Flavin-adeninedinucleotide in tissues of rats on diet deficient in flavin. Nature 144: 787.

Orr, J. B. 1936. Food, Health, and Income. London: Macmillan and Company.

—— 1941. Nutrition and human welfare. Nutrition Abstracts and Reviews 11: 3–11.

Orr, J. B., W. Thomson, and R. C. Garry. 1935. Long term experiment with rats on human dietary. J. Hyg. 35: 476–497.

Osborne, T. B., and L. B. Mendel. 1911. Feeding Experiments with Isolated Food Substances. Carnegie Institution of Washington, Publication 156, Parts I and II. Also papers in the Journal of Biological Chemistry, 1911–1924.

Osborne, T. B., and L. B. Mendel. 1926. The relation of the rate of growth to diet. J. Biol. Chem. 69: 661–673.

Park, E. A. 1923. Etiology of rickets. Physiol. Rev. 3: 106–163.

—— 1940. The therapy of rickets. J. Am. Med. Assoc. 115: 370–379.

Peters, J. P., and D. D. Van Slyke. 1931. Quantitative Clinical Chemistry: Interpretations. Baltimore: Williams and Wilkins.

Peters, R. A. 1936. The biochemical lesion in vitamin B_1 deficiency. Lancet 1936, I: 1161–1164.

Peterson, W. H., and J. T. Skinner. 1931. Distribution of manganese in foods. J. Nutrition 4: 419–426.

Platt, B. S., and G. D. Lu. 1939. The accumulation of pyruvic acid and other carbonyl compounds in beriberi and the effect of vitamin B_1. Biochem. J. 33: 1525–1537.

Poole, M. W., B. M. Hamil, T. B. Cooley, and I. G. Macy. 1937. Stabilizing effect of increased vitamin B_1 intake on growth and nutrition of infants. Am. J. Diseases Children 54: 726–749.

Roberts, L. J. 1935. Nutrition Work with Children. 2d edition. Chicago: University of Chicago Press.

Roberts, L. J., M. H. Brookes, et al. 1939. Supplementary value of the banana in institution diets. I, II. J. Pediatrics 15: 25–42, 43–53.

Rose, M. S. 1938. Foundations of Nutrition. 3rd edition. New York: The Macmillan Company.

—— 1939. Nutrition and the health of the school child. J. Am. Dietet. Assoc. 15: 63–85.

—— 1940. Feeding the Family. 4th edition. New York: The Macmillan Company.

Rose, M. S., and G. MacLeod. 1925. Maintenance values for the proteins of milk, meat, bread and milk, and soy-bean curd. J. Biol. Chem. 66: 847–867.

Rose, M. S., and E. L. McCollum. 1928. Growth, reproduction, and lactation on diets with different proportions of cereals and vegetables. J. Biol. Chem. 78: 535–547.

Rose, M. S., and E. McC. Vahlteich. 1938. A review of investigations on the nutritive value of eggs. J. Am. Dietet. Assoc. 14: 593–614.

Rose, W. C. 1938. The nutritive significance of the amino acids. Physiol. Rev. 18: 109–136.

Rous, P., and D. R. Drury. 1925. Outlying acidosis. J. Am. Med. Assoc. 85: 33–35.

Rowntree, J. I. 1930. A study of the absorption and retention of vitamin A in young children. J. Nutrition 3: 265–287.

Sandels, M. R., H. D. Cate, K. P. Wilkinson, and L. J. Graves. 1941. Follicular conjunctivitis in school children as an expression of vitamin A deficiency. Am. J. Diseases Children 62: 101–114.

Sandels, M. R., and E. Grady. 1932. Dietary practices in relation to the incidence of pellagra. Arch. Internal Med. 50: 362–372.

Schmidt, C. L. A., and D. M. Greenberg. 1935. Occurrence, transport, and regulation of calcium, magnesium, and phosphorus in the animal organism. Physiol. Rev. 15: 297–434.

Schoenheimer, R. 1942. The Dynamic State of Body Constituents. Cambridge, Mass.: Harvard University Press.

Schultze, M. O., E. Stotz, and C. G. King. 1938. Studies on the reduction of dehydroascorbic acid by guineapig tissues. J. Biol. Chem. 122: 395–406.

Sebrell, W. H. 1940. Nutritional diseases in the United States. J. Am. Med. Assoc. 115: 851–854.

Sheets, O. 1941. Conserving minerals and vitamins in vegetables. Mississippi Agricultural Experiment Station, Bull. 362.

Sherman, H. C. 1933. Food Products. 3d edition. New York: The Macmillan Company.

—— 1941. Chemistry of Food and Nutrition. 6th edition. New York: The Macmillan Company.

Sherman, H. C., and E. L. Batchelder. 1931. Further investigation of quantitative measurement of vitamin A values. J. Biol. Chem. 91: 505–511.

Sherman, H. C., and M. L. Cammack. 1926. A quantitative study of the storage of vitamin A. J. Biol. Chem. 68: 69–74.

Sherman, H. C., and H. L. Campbell. 1928, 1930. Relation of food to length of life. Proc. National Acad. Sci. 14: 852–855; J. Nutrition 2: 415–417.

—— 1937. Nutritional well-being and length of life as influenced by different enrichments of an already adequate diet. J. Nutrition 14: 609–620.

Sherman, H. C., H. L. Campbell, and C. S. Lanford. 1939. Experiments on the relation of nutrition to the composition of the body and the length of life. Proc. National Acad. Sci. 25: 16–20.

Sherman, H. C., and C. S. Lanford. 1943. Essentials of Nutrition. 2d edition. New York: The Macmillan Company.

Sherman, H. C., and F. L. MacLeod. 1925. The calcium content of the body in relation to age, growth, and food. J. Biol. Chem. 64: 429–459.

Sherman, H. C., and C. S. Pearson. 1942. Modern Bread from the Viewpoint of Nutrition. New York: The Macmillan Company.

Sherman, H. C., and E. N. Todhunter. 1934. The determination of vitamin A values by a method of single feedings. J. Nutrition 8: 347–356.

Shohl, A. T. 1939. Mineral Metabolism. New York: Reinhold Publishing Corporation.

Smith, A. H. 1931. Phenomena of retarded growth. J. Nutrition 4: 427–442.

Smith, M. C., E. M. Lantz, and H. V. Smith. 1931. The cause of mottled enamel, a defect of human teeth. Arizona Agricultural Experiment Station, Bull. 32.

Smith, M. C., and H. V. Smith. 1940. Observations on the durability of mottled teeth. Am. J. Public Health 30: 1050–1052.

Smith, M. C., and H. Spector. 1940. Further evidence of the mode of action of vitamin D. J. Nutrition 20: 197–202.

Smith, S. L. 1939. Vitamin needs of man: Vitamin C. U.S. Dept. Agriculture Year Book "Food and Life," pp. 235–255.

Speirs, M. 1939. The utilization of the calcium of various greens. J. Nutrition 11: 275–291.

Spies, T. D., W. B. Bean, R. W. Vilter, and N. E. Huff. 1940. Endemic riboflavin deficiency in infants and children. Am. J. Med. Sci. 200: 697–701.

Spies, T. D., A. P. Swain, and J. M. Grant. 1940. Clinically associated deficiency diseases. Am. J. Med. Sci. 200: 536–541.

Stearns, G. 1939. The mineral metabolism of normal infants. Physiol. Rev. 19: 415–438.

Stearns, G., P. C. Jeans, and V. Vandecar. 1936. Effect of vitamin D on linear growth in infancy. J. Pediat. 9: 1–10.

Steenbock, H., and A. Black. 1924. The induction of growth-promoting and calcifying properties in a ration by exposure to ultra-violet light. J. Biol. Chem. 61: 405–422.

Stiebeling, H. K. 1933. Food budgets for nutrition and production programs. U.S. Dept. Agriculture, Misc. Publ. 183.

—— 1941. Are We Well Fed? U.S. Dept. Agriculture, Misc. Publ. 430.

Stiebeling, H. K., and F. Clark. 1939. Planning for good nutrition. U.S. Dept. Agriculture Year Book "Food and Life," pp. 321–340.

Stiebeling, H. K., M. Farioletti, F. V. Waugh, and J. P. Cavin. 1939. Better nutrition as a national goal. U.S. Dept. Agriculture Year Book "Food and Life," pp. 380–402.

Stiebeling, H. K., and H. E. Munsell. 1932. Food supply and pellagra incidence in 73 South Carolina families. U.S. Dept. Agriculture, Tech. Bull. 333.

Stiebeling, H. K., and E. F. Phipard. 1939. Diets of families of employed wage earners and clerical workers in cities. U.S. Dept. Agriculture, Circ. 507.

Stotz, E., C. J. Harrer, M. O. Schultze, and C. G. King. 1938. The oxidation of ascorbic acid in the presence of guineapig liver. J. Biol. Chem. 122: 407–418.

Sydenstricker, V. P., W. H. Sebrell, H. M. Cleckley, and H. D.

Kruse. 1940. The ocular manifestations of ariboflavinosis. J. Am. Med. Assoc. 114: 2437-2445.

Taylor, C. M. 1942. Food Values in Shares and Weights. New York: The Macmillan Company.

Thorbjarnarson, T., and J. C. Drummond. 1938. Conditions influencing the storage of vitamin A in the liver. Biochem. J. 32: 5-9.

Todhunter, E. N. 1942. Evaluation of nutritional status. J. Am. Dietet. Assoc. 18: 79-82.

Todhunter, E. N., and R. C. Robbins. 1940. Observations on the amount of ascorbic acid required to maintain tissue saturation in normal adults. J. Nutrition 19: 263-270.

Todhunter, E. N., R. C. Robbins, and J. A. McIntosh. 1942. The rate of increase of blood plasma ascorbic acid after ingestion of ascorbic acid (vitamin C). J. Nutrition 23: 309-319.

Toepfer, E. W., and H. C. Sherman. 1936. The effect of liberal intakes of calcium or calcium and phosphorus on growth and body calcium. J. Biol. Chem. 115: 685-694.

Vahlteich, E. McC., E. H. Funnell, G. MacLeod, and M. S. Rose. 1936. Egg yolk and bran as sources of iron in the human dietary. J. Am. Dietet. Assoc. 11: 331-334.

Van Duyne, F. O., C. S. Lanford, E. W. Toepfer, and H. C. Sherman. 1941. Life-time experiments upon the problem of optimal calcium intake. J. Nutrition 21: 221-224.

Van Slyke, D. D. 1942. Physiology of the amino acids. Science 95: 259-263.

Wald, G. 1935. Carotenoids and the visual cycle. J. Gen. Physiol. 19: 351-371.

Weiss, S., and R. W. Wilkins. 1937. Disturbance of the cardiovascular system in nutritional deficiency. J. Am. Med. Assoc. 109: 786-793.

Wheeler, G. A. 1924. Pellagra in relation to milk supply in the household. Public Health Repts. 39: 2197-2199.

Whipple, G. H. 1935. Hemoglobin regeneration as influenced by diet and other factors: Nobel prize lecture. J. Am. Med. Assoc. 104: 791-793.

Whipple, G. H., and F. S. Robscheit-Robbins. 1940. Amino acids and hemoglobin production in anemia. J. Exptl. Med. 71: 569-583.

Wiehl, D. G. 1942. Diets of high school students of low income families in New York City. Milbank Memorial Fund Quarterly 20: 61-82.

Wiehl, D. G., and H. D. Kruse. 1941. Medical evaluation of nutritional status, V. Prevalence of deficiency diseases in their subclinical stage. Milbank Memorial Fund Quarterly 19: 241–251.

Wilder, R. M. 1941. Nutrition in the United States. A program for the present emergency and the future. Ann. Internal Med. 14: 2189–2198.

—— 1942. Nutrition and national defense. J. Am. Dietet. Assoc. 18: 1–9.

Williams, R. J., and R. T. Major. 1940. Structure of pantothenic acid. Science 91: 246.

Williams, R. R., and T. D. Spies. 1938. Vitamin B_1 (Thiamin) and Its Use in Medicine. New York: The Macmillan Company.

Winters, J. C., A. H. Smith, and L. B. Mendel. 1927. The effects of dietary deficiencies on the growth of certain body systems and organs. Am. J. Physiol. 80: 576–593.

Yavorsky, M., P. Almaden, and C. G. King. 1934. The vitamin C content of human tissue. J. Biol. Chem. 106: 525–529.

Youmans, J. B., and M. B. Corlette. 1938. Specific dermatoses due to vitamin A deficiency. Am. J. Med. Sci. 195: 644–650.

INDEX

Adrenine, 84

Adulthood: Age, *see* Maturity

Agricultural Experiment Station at Cornell, 141

Agricultural production, effect of consumer demand upon, 176-81

Agriculture, Department of, 24, 166; classification of food, 94, 109, 112; Yearbook and other publications, 116, 127, 153, 163, 165, 166, 170, 172, 174, 178, 179; Home Economics Bureau's food groups and meal plans, 153, 154; milk-for-health campaigns, 179

American Medical Association, 70; *Journal*, 74, 128, 191*n*

Amino acids, constituents of proteins, 22-26, 83, 84; differences, 24; dynamic relationships in field of, 89; interchanges, 90

Ammonia, 89, 90

Anemias and their causes, 31-33; types, 32

Animal experimentation, accuracy of nutritional studies, 5; why a quantitative research instrument, 7; studies of farm animals, 13, 141; why rats selected for nutritional experiments, 28*n*, 188; relation of calcium content of food and of body, 29; rewarded by discovery of vitamins, 37, 38; function served, 137; results significant in human experience, 140, 188; feeding experiments at Columbia University, *see* Protective foods

Antiberiberi vitamin, 59; *see* Vitamin B

Antineuritic substance, 39, 58

Antirachitics, 80

Antiscorbutic vitamin, 45; *see* Vitamin C

Appetite stabilization, 60

Apples, 111

"Are the Days of Creation Ended" (Merriam), 147

Ariboflavinosis, 69

Armsby, H. P., 11, 12

Army Medical Commission, 39, 58, 59

Ascorbic acid, 48; *see* Vitamin C

Astor, Lord, 130, 170

Atwater, W. O., 10, 11, 12, 24

Autolysis, study of, 87

Babies, calcium content at birth, 201; *see also* Children

"Backward Art of Spending Money, The" (Mitchell), 149

Bananas, 111

Basal metabolism tests, 13

Batchelder, E. L., 76, 216

Beef, *see* Meat

Benedict, F. G., 10, 11, 12, 13, 19, 24

Beriberi, 39, 40, 57-60

Bernard, Claude, 4

Bessey, O. A., 70

Bibliography, 229-42

Blacktongue, 66, 67

Blatherwick, N. R., 110

Blood, importance of sodium to, 27; of calcium, 27, 30; of iron, 30, 31; anemias, 31-33; stabilization of calcium content of, 211 f.

Blyth, A. W., 67

Bodansky, O., 76

Body, as mechanism: food as fuel, 9-21, 82, 87; adjustment to different levels of energy exchange, 19; advances in knowledge of materials of, 22-36 (*see* Materials); idea of a *fixité* of internal environment, 52; how nutritional resources managed by, 82-91; steady states, 86; dynamic equilibrium states, 86 ff.; regulator of imbalance conditions in, 207

Bomb calorimeter, 10

Bone trabeculae, 29, 210-13

Bei Fragen zur Produktsicherheit wenden Sie sich bitte an:
If you have any questions regarding product safety,
please contact:

Walter de Gruyter GmbH
Genthiner Straße 13
10785 Berlin
productsafety@degruyterbrill.com